SEAWEED
and PLANT GROWTH

T.L. SENN, Ph.D.

(ASCOPHYLLUM nodosum)
is a specie of seaweed which grows in the littoral zone,
i.e., the strip of coastline which is left dry at low tide.

First Edition — 1987

Edited by
T.L. (Tee) Senn, B.S., M.S., HMS, Ph.D.
Head and Professor of Horticulture Emeritus
Clemson University
Clemson, SC 29634-0345 U.S.A.

Copyright © 1987 by T.L. Senn. All rights reserved.

Neither this book nor any part may be reproduced or transmitted in any form by any means, electronic or mechanical, including photocopying, microfilming, and recording, or by any information storage and retrieval system, without permission from the author.

Library of Congress Catalog Number: 87-90524
ISBN: 0-939241-01-3

Printed in the United States of America.

Preface

Dr. T.L. Senn, Head Professor Emeritus, Department of Horticulture, Clemson University (known to his friends as Tee), has been involved in seaweed research for horticulture and agriculture for almost 30 years. This research has been funded with both private and federal grants.

Tee has published numerous papers on his research and his research is known all over the world. He is without doubt one of the leading seaweed researchers in the world.

Unfortunately, many of his reports and publications have been used out of context by both individuals and companies with no relationship to his research work. This has given seaweed a bad reputation, and to a large extent, caused problems for companies serious about marketing seaweed products properly.

Tee has an impeccable academic background and has written this book in a language easily understood by most gardeners, growers, and farmers. The data presented in the book are properly documented and will be of invaluable assistance to companies planning to use seaweed in their fertilizer and spray formulations. I am convinced that this book will bring seaweed out of the "snake oil" age and that the regulatory agencies will look at seaweed from a different viewpoint.

I know of few products that have been more criticized

than seaweed. On the other hand I know of no other individual who possesses the patience and the quiet persistence in research with seaweed as Tee. He is a seeker of truth, his research is unequaled and all his works reflect these facts.

Tee loves life, he loves nature, he loves people and this is reflected in all his actions both in his professional and personal life. He developed a 40-acre garden for the blind at Clemson University. This garden is visited by thousands of visitors every year and is known nationwide. He has developed programs to employ physically handicapped persons and other less fortunate in our society.

The timing for publishing this book is perfect. We all face serious problems with air, water and soil pollution and this book provides the readers with documented advice on the use of this nonpoisonous and nonpollutant alternative — seaweed.

I value very highly the friendship that I have developed with Tee over the last 30 years. Little did I know when I arrived from Norway that the marketing of Norwegian **ASCOPHYLLUM nodosum** seaweed in the United States would be that difficult, yet fascinating and extremely interesting. The efforts would have been futile without Tee's research and valuable support. I have many great memories from our travels here and abroad to seaweed symposia and seminars and meeting with numerous government officials. I feel very fortunate to be part of this association and the continuing effort to expand the knowledge of the use of seaweed.

Per Bye Ohrstrom

About the Author
by Kelly Winters
Clemson World, April 1986

Flowers of the Sea
"Call us not weeds, we are flowers of the sea."
                    E.L. Aveline

Taze Leonard Senn . . . known as Tee to his friends . . . collects quotes as a hobby — using them for inspiration and as a way to express himself.

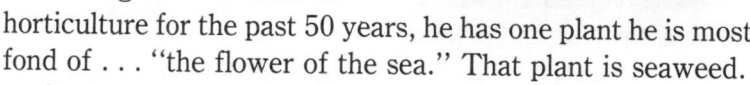

Having been a man of horticulture for the past 50 years, he has one plant he is most fond of . . . "the flower of the sea." That plant is seaweed.

Seaweed — what good is that slimy stuff, you say? Dr. Senn has a very good answer.

Tee Senn, born in Newberry, had planned to attend Newberry College, but was fortunate enough to receive a scholarship to Clemson. He came here in the fall of 1935 with plans to major in dairying.

"The name 'Senn' means keeper of the cow," Tee explained. "Most of my family members were dairymen or teachers, so that was how I decided I'd major in dairying." But that decision turned out to be short-lived.

A job he got his first day on campus changed all that. He went to work in the greenhouse of the University's horticulture department. "I had never heard of horticulture," he recalls . . . "but on that first day, I had a job with them, cutting castor beans and okra. It changed my life, that job." After switching his major in his sophomore year from dairying to horticulture, Senn's interest in his new field continued to grow.

After graduating, Senn became a horticulture assistant at Clemson — a research position that got him interested in the research angle of looking at plants and their lives. But he was not able to scratch his "research itch" for several years.

After teaching briefly at Clemson, Tee went to the Pacific, where he served as a Navy communications officer in World War II. Several years after he returned to the campus, he received his master's degree in horticultural physiology from the University of Maryland, while on a Dean's Fellowship; later he received his Ph.D. in the same subject there, in 1958. That same year he received a Danforth Teaching Fellowship — the only South Carolinian in agriculture to ever receive one — a fact he is still proud of today.

While on the Fellowship, the result of a nomination by the head of the horticulture department, he was able to do research on controlled atmospheric storage. "I was able to regulate plants and their growth by manipulating what I gave them," he explains.

He notes that this got him wondering what could be done to improve soil . . . what could you add to it to make it better, to help it aid plants in their growth process?

About the same time that Senn's interest in growth promoting substances peaked, he met a Norwegian who told him about the growth-promoting properties of seaweed. After talking and realizing that they had stumbled upon an idea that needed to be researched, Tee went to the National

Science Foundation in Washington, D.C., to ask for a grant to pursue that objective.

When he rose to become head of the horticulture department in 1960, Tee boldly told the NSF about his plans. "I went to the aristocracy of agriculture," he pointed out, "because when you mention seaweed to just anyone, they laugh and turn up their noses!"

Luckily, the NSF did neither. After Senn drew up a proposed project outline, based on analysis and exploration of seaweed, he got the grant he needed for preliminary research.

After getting more research grants — including one that enabled him to collect seaweed from all over the world — Tee was happy to learn that the area of seaweed research was eventually recognized as a legitimate practice . . . thanks in no small part to Tee Senn.

He stressed that only certain kinds of seaweed can be used for certain things, but that all of us use seaweed every day in various ways. "Everyone uses it subconsciously," he observed, rattling off a list of products that contain the plant as an ingredient: lipstick, shampoo, ice cream, beer, and plastic, plus many others. He continued to promote seaweed in every way he knew how, even after retiring from the university in June 1981.

Today he travels all over the world . . . to the Middle and Far East, Europe, Central and South America and many states in the U.S.A. — giving lectures about seaweed, designing research projects for companies and manufacturers, working on new seaweed product development and serving as a consultant for companies negotiating with each other about joint seaweed projects.

Senn cautioned that he does not imply that seaweed can do everything . . . pointing to a picture of a balding man he's done business with, he exclaimed, "It certainly doesn't grow hair!"

Other plants besides seaweed have attracted his interest and he has expressed that interest in various ways, helping others at the same time. Senn had a Clemson-produced radio program called "The Plant Professor" back in the 1960's which eventually aired on 50 stations. He talked about horticulture and general plant problems, as he did on the "At Home" show, a Clemson television series broadcast in the same decade.

Tee thought about the people he wasn't reaching through these two programs, and began to wonder how he could tell them about his great love: plants. Realizing that he was not reaching the blind, Senn went to the S.C. School for the Blind and Deaf in Spartanburg, inquiring how he could convey the beauty of plants to them.

He collected many ideas and came to realize, with the help of a child from the school named Ricky, that there were many different ways a blind person could distinguish between plants ... by smell, by leaf patterns, by following a Braille trail, etc.

Through a grant, Tee was able to hire several people to work on the concept of a garden for the blind; he took his team and the idea to officials participating in National Lawn and Garden Week in Washington. His suggestion — that blind people be allowed to participate in this occasion through a garden made especially for them — was received with enthusiasm.

When he returned to Clemson it was decided that the university should have its own garden for the blind. After a lot of work by Tee and his team, that garden opened in 1970, and has drawn thousands of visitors from the U.S. and overseas to the campus ever since.

As he contemplated retirement, another of his favorite quotes — this one from Dryden — came to mind:

A foundation of good sense and a cultivation of learning are required to give a seasoning to retirement and make us taste its blessings.

Tee Senn emphatically stated that one thing he did not retire to do was to sit down. The only time he sits now, it seems, is for an interview. He is president of Clemson's Kiwanis Club and a charter member of Clemson's Kiwanis' Golden K. Club, a group of 44 retirees. He's a member of the Elk's Club and has been its state scholarship chairman for the past 15 years — aiding students with plans to continue their education through college.

Senn is permanent president, secretary and treasurer of Clemson's Class of 1939 — a class that is now trying to raise half a million dollars to be donated to the university in 1989, Clemson's Centennial. That year will also be the class' 50th anniversary . . . and the 50th wedding anniversary of Tee and his wife, Marguerite.

His future plans include finishing a book for growers, distributors and dealers of the horticulture world, titled **Seaweed and Plant Growth.** It will be his second literary achievement . . . as he published another book about plants, **The Fundamentals of Horticulture,** a venture he coauthored with two others and which is now available in six languages.

Tee Senn believes that teaching at Clemson was one of the most rewarding experiences of his life. "I believe in teaching very much," he said. "I think it's one of the greatest professions on earth." He added with a smile, "I do detest the word 'professor' though . . . I prefer teacher. A teacher teaches, a professor professes."

His personal teaching methods have been recognized in many ways. Besides experiencing the satisfaction of unveiling the wonders of the world of horticulture to his students, he was awarded both the Southern Region and National L.M.

Ware Distinguished Teaching Award in 1968.

Despite all his travels, he still calls Clemson home — relaxing in his spare time in his large home greenhouse by the lake, and collecting old memorabilia such as music boxes and player pianos. Home base will always be Tiger country for the roots of Tee Senn.

"Clemson is my love. I don't think I'd want to live anywhere else."

Professional Societies and Honors

South Carolina Horticulture Society
American Society for Horticultural Science
Gamma Sigma Delta-Sigma Xi-Alpha Zeta-Pi Alpha Xi-Gamma Rho Fraternity
Life Membership — Garden Club of South Carolina and Garden Club of America
American Society for Testing Materials
Special Award — Garden Club of South Carolina
American Men of Science
Leaders in Science — Outstanding Educator of America
Who's Who in South and Southwest
Who's Who in American Education
Fellow — American Association for Advancement of Science
Recipient of Outstanding Horticultural Teaching Award for Southern Region — 1968
Recipient of National Outstanding Horticultural Teaching Award — 1968
Fellow — South Carolina Horticulture Society — 1984
Recipient of the National Council of Garden Clubs Silver Seal Award — May 1984 (for achievements in hortitherapy)
Received Clemson University Distinguished Alumnus Award — 1974
Elk of the Year — 1979 — Scholarship State Chairman for 15 years
Received Presidential Citation from the Garden Club of

South Carolina — 1981

Received Gamma Sigma Delta Distinguished Service Award — 1981

Received Distinguished Service Award presented by peach growers of South Carolina — 1981

Received certification as Master Horticultural Therapist from NCTRH — 1981

Special Award — South Carolina Turf and Ornamentals Industry

Special Award — South Carolina Foundation Seed Association

Special Award — South Carolina Seedsmen Association

Has consulted in agricultural affairs in Canada, Iran, Kuwait, Norway, Puerto Rico, Costa Rica, Spain and Pakistan

Also has traveled widely in Europe, Australia, New Zealand, Central and South America, China, Spain, Portugal, and the Philippines

Publications

Over 100, of which 20 relate to seaweed research.

Acknowledgements

The author expresses his appreciation to Per Bye Ohrstrom for his time and effort, in so many ways, that have helped to make this manuscript a reality. To all the growers, dealers, distributors, and processors who encouraged me to write this publication, I offer my thanks. These individuals are the true believers in the beneficial effects of seaweed products. My special thanks to my daughter-in-law Cleve Ann Senn for her patience and understanding during the excellent typing of the manuscript. My thanks to the Senn boys — Dick, Tom, Dave, Richie, and Shane — for listening to the seaweed story over and over for many years.

To my wife, Marguerite (Reet),
loyal supporter, confidant, my friend,
my lifelong companion.

T.L. Senn

To the Reader

Seaweed, no doubt, outdates man by a considerable period of time. Man has been using seaweed for food and in the growth of plants since his very beginning. Early man, perhaps, cared less why seaweed influenced plant growth. He found them, used them and was happy.

There are numerous reports indicating that seaweed extracts are responsible for increased crop yields. The reasons why seaweed extracts are beneficial to plant growth are still not fully understood. This is not so difficult to appreciate, for anyone working with plants knows that "nature reveals her secrets reluctantly."

These new seaweed products that are available to the plant grower today are the result of many years of research and development by many individuals.

Unfortunately, this new group of seaweed products has been misunderstood and misrepresented. As you read this publication, it is requested that you keep an open mind and evaluate the merits of the information as it relates to your welfare.

> T.L. (Tee) Senn, Dr.
> 201 Strawberry Lane
> Clemson, South Carolina U.S.A. 29631

Contents

Chapter 1 Introduction 1-1
    Alternative Methods of Crop Production
    Different Kinds of Seaweeds   Utilization
    of Seaweed   International Seaweed
    Symposia   Seaweed Research Publications

Chapter 2 Marine Algae-Seaweed 2-1
    Classification   Composition   Technical Data

Chapter 3 Facts to Build On 3-1
    Literature Review   Micronutrients
    PGR   Literature Cited

Chapter 4 What Makes Plants Grow 4-1
    Plant Processes   Plant Growth   Seed   Root
    Stem   Leaf   Flower   Growth Rates

Chapter 5 Plant Growth and Development 5-1
    Definitions   Hormones and Micronutrients
    Seaweed Promotes Plant Growth and Development
    Selected References

Chapter 6   How Plants Mature.....................6-1

   The Vegetative Phase Growth
   The Reproductive Phase Maturity   Essential
   Elements Limiting Factor   Mature Maturation
   Maturity Ripe   Good Farming Practices

Chapter 7   Why Plants Need Micronutrients..........7-1

   Biochemical Reactions in Crop Plants
   Essential Elements   Essential Raw Materials
   Micronutrients   Role of Micronutrients
   Selected References

Chapter 8   Plant Growth Regulators 8-1

   Classification   Enzymes   Hormones
   Practical Applications   Bioassays   Gibberellin
   Auxins   Cytokinins   Research Reports
   Future Outlook

Chapter 9   Seaweed and Plant Stress................9-1

   Stress   Plant Stress   Seed Physiology
   Flowering   Environmental Stress   Flowering
   Plant Response to Stress   References

Chapter 10 They Condemn What They
                  Do Not Understand 10-1

   Biological Farming   Basics   Development
   Adverse Conditions   Micronutrients
   How Could It Happen?
      Seed   Root Growth   Flowering
      Cold Hardiness   Insects   Nematodes
      Shelf Life
   The Answers   Grower Survey   Points to
   Ponder   Summary   Selected References

Chapter 11 Speak the Language 11-1

    Bioliving   Biochemical   Biophysical
    Essential Elements   Chelation   Seaweed
    Contains Growth Regulators   Plant Foods
    ppm-ppb   Language Communicate Understand

Chapter 12 There Is a Time for All Things 12-1

    Time   Day   Disease   Things to Consider
    Activities   Concentration   Beware
    The Team   Time and Research March On

CHAPTER 1
Introduction

The sea never changes and its works, for all the talk of men, are wrapped in mystery (Joseph Conrad, Typhoon).

Man has always been an explorer. He is interested in the regions above and below his earth, including the oceans and seas. Man of necessity is concerned about the environment in which he lives. This environment is living and active and is the home for many living things in the air, the soil and the water. The bacteria, fungi, and nematodes represent the lower forms of life, whereas many kinds of plants on land and in the oceans represent the higher forms of living things. Current interest in alternative methods of crop production from growers, educators, environmentalists, and a cross section of the general population prompts this presentation of information concerning marine algae-seaweed for agricultural and horticultural usage.

When you prepare your favorite tossed salad at home or at your local salad bar, you may be getting more than tomatoes, cucumbers, lettuce, and other vegetables. You could be dishing up a helping of harmful residues. It is estimated that more than half the food we eat contains applied chemicals, many of them are residues from the 2.3 billion pounds of pesticides farmers apply to their crops each year. Some pesticides are suspected carcinogens; others are highly toxic. According to various chemical manufacturers, these pesticide chemicals are safe for human consumption when farmers apply them according to directions. A lot of growers still believe "if a little bit is good, add some more, and it will be better." Research has shown the applications of Norwegian **ASCOPHYLLUM nodosum** extracts to various fruits and vegetables prolongs their shelf life and no preservatives at the salad bar are needed.

In a recent publication, *Introduction to Insect Management,* Dr. Robert Metcalf, world's authority on insects, stated that "farmers use more than twice as much pesticides as they need and still get poor results." He further stated, "We now suffer about 20% crop loss to insects, the same as we did in 1900." Dr. Metcalf also said, "We've won some individual battles along the way, but we're losing the war." Dr. Lowell N. Lewis, Director of the Agricultural Experiment Station, University of California, states that:

> *The real issue facing plant science is that of the impact that producing our food has on the quality of our environment. A recent report from the U.S. Department of the Interior indicates that, of all the toxic problems associated with natural systems, almost half were caused by agricultural pollution. The only choice is to move forward as rapidly as possible to increase our basic knowledge of plant sciences that will get us off the treadmill of mounting environmental problems.*

Two-thirds of the earth's surface is covered with water, and marine algae constitutes most of the whole vegetation which exists in this area. Algae derive their nourishment from substances held in solution by water. The water in which commercial seaweed grows is non-polluted and therefore the products are non-toxic and safe for use on food crops. They do not contaminate the environment, thereby helping man to maintain a healthy world in which to live.

The number of different kinds of seaweed found throughout the world is probably astronomical. However, only a few seaweed orders are harvested for agricultural usage; namely LAMINARIACEAE, FUCACEAE, and GIGARTINACEAE. In the research literature these marine algae are usually designated as **ASCOPHYLLUM nodosum** (PHAEOPHYTA) L., **DURVILLEA potatorum** (LABILLARDIERE) Areshoug, and **ECKLONIA maxima** (Osbeck) Papenf.

Utilization of seaweed in crop production dates back to antiquity. With the development of chemical nutrients in the late 1800's, the use of natural products in crop production began to decrease in popularity. In recent years when adverse effects of the addition of many chemicals to the environment were questioned, natural sources of plant nutrients and plant growth promoters have become popular and in wide usage. Commercial seaweed extracts became available in 1950 — the most commonly used marine algae being **ASCOPHYLLUM nodosum** order **FUCACEAE**. Seaweed extracts are generally applied as foliar sprays. When foliar feeding became accepted as orthodox practice in the 1960's, this notable change boosted the interest in liquid seaweed products throughout the world. It was at this time that the author became interested in the potentials of seaweed extracts in agriculture, and initiated his first research projects with Norwegian **ASCOPHYLLUM nodosum.**

Plant growth and development is controlled by plant growth regulators (hormones) produced by the plant itself. Many synthetic plant growth promoting substances are now being widely used in crop production. As early as the 1950's scientists, including the author, reported evidence suggesting the presence of growth hormone(s) of some type in seaweed extracts. The growth promoting substances in marine algae reported include IAA, gibberellin, cytokinin, and substances of a similar composition. The amount and type of plant growth regulators which exist in seaweed products have been widely discussed. The problem was associated with inadequate methods for determinations of these substances. Responses to various bioassays were positive but no quantitative methods were known until the early 1980's. Presently there are numerous, highly satisfactory methods described and used for gas liquid chromatographic (GLC) determinations of these plant growth regulating substances in crude aqueous solutions of several marine algae.

Scientists are continually seeking a basic understanding through research, and seek applications of sea plants for meeting man's needs for survival.

An international seaweed symposia was started in response to growing worldwide interest in the commercial potential of seaweed-derived gelling and emulsifying agents, which led to the organization of a conference held in Halifax, Canada, in 1948. This conference was attended by Dr. F.N. Woodward from Edinburgh, Scotland, who then organized the first international seaweed symposium held in Edinburgh, Scotland, in 1952. The purpose of these symposia was to further the understanding and utilization of seaweeds. Following the symposium in 1952, additional symposia were held in Trondheim, Norway (1953); Galway, Ireland (1958); Biarritz, France (1961); Halifax, Canada (1965); Santiago de Compostela, Spain (1968); Sapporo, Japan (1971); Banger, Wales (1974); Santa Barbara, California (1977); Gøteborg, Sweden (1980); China (1983); and in Sao, Brazil (1986).

The ability to cope with worldwide food and dwindling resource problems of the future may well reside in symposia such as the ones mentioned above. The research reported contributes to the advancement of knowledge related to seaweed and its practical applications in such areas as food production, energy resources, and man's health. These gatherings of people from academia, industry, and government, representing interests in research and commercialization, have been of mutual benefit and have aroused in the public a large and new awareness of the importance that sea plants hold for mankind.

The proceedings of these symposia have been cited frequently in this publication. In 1958, the author became concerned about the lack of information available on plant growth substances and in general "how plants grow." It was his good fortune to be introduced to the research being conducted with seaweed by the Norwegian Institute for Seaweed Research. Since that date he has been actively involved with seaweed research at Clemson University, has attended international seaweed symposia, published many popular and scientific articles, and has traveled extensively throughout the world learning more about the uses of seaweed in agriculture and horticulture. He was fortunate to receive National Science Foundation, National Oceanic and Atmospheric Administration, Economic Development Administration, and industry grants to support his research. He has had the pleasure of meeting and getting to know the outstanding leaders in the international seaweed industry. A listing of seaweed publications resulting from his research at Clemson University is found at the end of this chapter.

A number of highly respected individuals have talked seriously with the author about the seaweed industry — its image, its future, and the vital need for unity and cooperative research. They feel, and he concurs, that it is now time for the leaders to get together and discuss the issues related to the uses of seaweed (kelp) in high technology agriculture and horticulture.

Pertinent issues that need addressing as we move to wider uses of seaweed include: regulatory, research, labels and integrity. It is essential that we develop standards for the industry which can only enhance the credibility and integrity of seaweed as an agricultural and horticultural production tool. It is important that the leaders in seaweed usage establish these standards rather than have the government set them for industry.

The presentation of research findings, not only of our own research but of well-known scientists throughout the world, should help roll back the wall of ignorance that surrounds the use of marine algae-seaweed in agriculture and horticulture. A group of interested and dedicated research and industry representatives in seaweed product agricultural uses met at Clemson University, Clemson, South Carolina, U.S.A., on March 6-7, 1986. The main purpose was to organize a kelp association to set standards of excellence for the seaweed agricultural use industry.

This publication is intended for general circulation and is written in such a non-technical language that it will enable those readers who are unaccustomed to the technical language to obtain an overview of the subject matter of the entire publication.

Seaweed Publications

1. Senn, T.L., J.A. Martin, J.H. Crawford, B.J. Skelton. February 1960. The effect of kelp meal on development and comparison of various vegetables and special crops. Southern Ag. Workers 57th Annual Report. p. 182.

2. Fox, D. et al., 1960. The effect of seaweed on plant growth. 57th Annual Meeting Amer. Soc. Hort. Sci., Oklahoma State Univ., Stillwater, Oklahoma.

3. Senn, T.L. and J.A. Martin. October 1960. Growth regulator action of seaweed. Departmental Memo.

4 Senn, T.L. and J.A. Martin, J.H. Crawford and C.W. Derting. September 1961. The effect of Norwegian seaweed (AS-

COPHYLLUM nodosum) on the development and composition of certain horticultural and special crops. S.C. Agr. Exp. Sta. Hort. Res. Ser. No. 23.

5. Aitken, J.B., T.L. Senn and J.A. Martin. September 1961. The effect of varying concentrations of Norwegian seaweed (ASCOPHYLLUM nodosum) on Duncan grapefruit and pineapple orange seedlings grown under greenhouse conditions. S.C. Agr. Exp. Sta. Hort. Res. Ser. No. 24.

6. Gambrell, C.W., E.T. Sims, Jr., R.D. Suber and Danny O. Ezell. February 1963. The effect of pre-harvest sprays of liquid seaweed extract, verdan and maleic hydrazide on storage and some quality indices of Sullivan Elberta peaches before and after storage. S.C. Agr. Exp. Sta. Hort. Res. Ser. No. 40.

7. Aitken, J.B. and T.L. Senn. February 1964. The effects of seaweed extract and humic acids on the O_2 uptake of **citrus sinensis** seedlings grown in nutrient element deficient culture. S.C. Agr. Exp. Sta. Hort. Res. Ser. No. 53.

8. Senn, T.L. and B.J. Skelton. February 1966. Review of seaweed research — 1958-1965. S.C. Agr. Exp. Sta. Hort. Res. Ser. No. 76.

9. Skelton, B.J. and T.L. Senn. October 1966. Effect of seaweed on quality and shelf life of Harvest Gold and Jerseyland peaches. S.C. Agr. Exp. Sta. Hort. Res. Ser. No. 86.

10. Skelton, B.J. and T.L. Senn. 1968. Effect of seaweed on quality and shelf life of peaches. Proc. VI International Seaweed Symposium. Santiago de Compostela, Spain. pp. 731-736.

11. Senn, T.L. and B.J. Skelton. 1968. The effect of Norwegian seaweed on metabolic activity of certain plants. Proc. VI International Seaweed Symposium, Santiago de Compostela, Spain. pp. 723-730.

12. Senn, T.L., J.P. Fulmer, and B.J. Skelton. 1970. Camellias at Clemson. Carolina Camellias, Vol. XXII. 7-11.

13. Senn, T.L. and B.J. Skelton. May 1972. Seaweed research at Clemson University. 1961-1971. S.C. Agr. Exp. Sta. Hort Res. Ser. No. 141.

14. Kingman, Alta R. and T.L. Senn. March 1973. A survey of seaweed research. S.C. Agr. Exp. Sta. Hort. Res. Ser. No. 146.

15. Kingman, A.R. and J.A. Lewis, III. July 1975. Effects of **ASCOPHYLLUM nodosum** extracts and meal on insect and mite populations and plant quality of several plants in the greenhouse. S.C. Agr. Exp. Sta. Hort. Res. Ser. No. 158.

16. Kingman, A.R. May 1975. Plant growth responses to extracts of **ASCOPHYLLUM nodosum**. S.C. Agr. Exp. Sta. Hort. Res. Ser. No. 161.

17. Senn, T.L. and A.R. Kingman. December 1975. A report of seaweed research, 1974-1975. S.C. Agr. Exp. Sta. Hort. Res. Ser. No. 162.

18. Senn, T.L. and A.R. Kingman. December 1975. A report of Erth-Rite research, 1974-1975. S.C. Agr. Exp. Sta. Hort. Res. Ser. No. 163.

19. Senn, T.L. and A.R. Kingman. December 1975. A report of Ferma-Lizer research, 1974-1975. S.C. Agr. Exp. Sta. Hort. Res. Ser. No. 164.

20. Senn, T.L. and A.R. Kingman. September 1978. Seaweed research in crop production, 1958-1978. Prepared for Economic Development Administration of the U.S. Dept. of Commerce. Project No. 99-6-09329-2.

21. Kingman, A.R. and T.L. Senn. 1977. Bioassay systems to test for plant growth hormones in extracts of **ASCOPHYLLUM nodosum**. J. Phycol. 13, p. 23.

22. Kingman, A.R. and J. Moore. 1982. Isolation purification and quantitation of several growth regulating substances in **ASCOPHYLLUM nodosum** PHAEOPHYTA. March 25 (4), 1982. pp. 149-154.

CHAPTER 2
Marine Algae-Seaweed

Call us not weeds; we are flowers of the sea (E.L. Aveline, *The Mother's Fables*).

The distinction between animal life and vegetable life (plants) is based on the way or manner of assimilating food. Plants take in mineral substances, referred to as assimilating inorganic matter. Animals require vegetable or some other organic matter for their subsistence.

Plants having no distinction of leaf or stem are called thallophytes. Algae are thallophytes that live in the water and receive their nourishment from the water. Algae reproduce by vegetative reproduction, breaking into more individuals, and by the production of spores which produce new plants on germination.

Marine algae are classified according to their differences, such as color — blue-green, grass-green, olive-green, brown, and red. They may also be classified according to their similarities or resemblances.

This classification is arranged as follows: kingdom, phylum, class, order, family, genus, and species. For example, one of the most widely researched seaweed for agricultural purposes would be Class — **algae,** Order — **FUCACEAE,** Genera — **ASCOPHYLLUM,** Species — **nodosum,** and is referred to as **ASCOPHYLLUM nodosum** or simply ASCOPHYLLUM. There are many other seaweed products made from **LAMINARIA digitata, MACROCYSTIS pyrifera, ECKLONIA maxima, DURVILLEA potatorum** and **CHONDRUS crispus.**

The chemical composition of seaweed is mainly determined from the conditions in which it grew. The temperature of the water and the amount of sunlight determine to a large extent the particular seaweed that will grow in a certain area. Brown seaweed grow more prolific in temperate waters, and red seaweed grow better in warmer waters. The amount of sunlight available influences the depth at which various seaweed will grow. The littoral shore zone covers the space between high tide and low tide. Seaweed growing in this zone is subjected to the sun and air, and then to complete submergence at varying periods. The **Fucus** grow very well in this zone. Another zone, laminarian, extends from low tide to a depth of about 90 feet. The **LAMINARIA** are prolific in this shore zone.

The composition of any one type of seaweed is always very consistent. Any variation in composition in any one type of seaweed is usually a seasonal variation. Typical technical data information of several seaweeds are listed below. These were prepared by the manufacturer and are not to be considered as an endorsement by the author.

ECKLONIA maxima (Osbeck) Papenf.

A concentrate is prepared from the brown alga **ECKLONIA maxima** (Osbeck) Papenf. by a cell burst process which does not involve the use of heat, chemicals or dehydration (Featonby-Smith and Van Staden, 1983).

Norwegian ASCOPHYLLUM nodosum
(L.) Lejol Marine Algae

INTRODUCTION:

A soluble powder and liquid concentrate is prepared by alkaline hydrolysis of Norwegian seaweed, **ASCOPHYLLUM nodosum** (L.) Lejol. Plant growth regulators are materials which control the growth and development of plants. These substances are present in small quantities, generally in range of parts per million (ppm) or even in parts per billion (ppb).

The main growth regulators which promote plant growth are the auxins, hormones, indoles and cytokinins.

Cytokinins are cell division factors which were first discovered in rapidly dividing cells during the 1950's. The indole compounds are naturally occurring plant growth promoters for root development and bud initiation.

The greater proportion of scientific research projects and crop response has been with **ASCOPHYLLUM nodosum.** The Norwegian **ASCOPHYLLUM nodosum** is to the author's knowledge the most widely researched type and his research has been confined mainly to this type. This is largely because the supply of the product is stable and the processing of the seaweed biomass is subject to rigid quality control.

GROWTH PROMOTERS:

The major plant growth promoters in **ASCOPHYLLUM nodosum** are almost certainly cytokinins. This is supported by bioassays and chemical analyses including gas liquid chromatography research. Adenine and Zeatin are the major cytokinins present in **ASCOPHYLLUM nodosum**. Zeatin is the most biologically active cytokinin known. Adenine exhibits lower activity. Gas liquid chromatographic analysis suggests that adenine (6-aminopurine) 6-(Y-Y-dimethyl-ally) (amino) purine, kinetin (6-furfurylaminopurine) and zeatin and 6-benzylaminopurine are present.

The **ASCOPHYLLUM nodosum** powder has an equivalent kinetin level of not less than 50 µg/g of dry powder.

Compound	Concentration of (q/15 ml H_2O)
Abscissic Acid (ABA)	0.02[z]
Adenine	0.03[z]
Indole Acetic Acid (IAA)	0.05[y]

z Based on 1 g dried **ASCOPHYLLUM nodosum** concentrates in 15 ml H_2O following extraction with ethyl acetate: pH 7.7

y Based on 1 g dried **ASCOPHYLLUM nodosum** concentrates in 15 ml H_2O following methanol extraction and partitioning with methylene chloride: pH 7.7

Gas liquid chromatographic retention times for several solutions of growth regulating substances extracted from **ASCOPHYLLUM nodosum**.

Compound	Retention time (seconds)
Adenine[z]	65-100
Adenosine[z]	325-340
Zeatin[z]	275-285
2-IP[z]	157-172
IAA[y]	85-112
ABA[z]	180-195

z Extraction with ethyl acetate.
y Methanol extraction followed by partitioning with methylene chloride.

SUGARS AND ORGANIC ACIDS:

Carbohydrates contain the elements carbon, hydrogen, and oxygen, usually with the carbon atoms combined with hydrogen and oxygen in the same proportion that hydrogen and oxygen exist in water. Examples are $C_6H_{12}O_6$ and $C_{12}H_{22}O_{11}$, glucose sugars manufactured by land plants. These provide the building blocks needed to make the storage forms of carbohydrates — sucrose, dextrin, and starch. For example, the grape stores large quantities of glucose in its fruit, sugar cane and sugar beets store large quantities of sucrose in their stems and roots respectively.

Seaplants produce mannose sugar and a sugar alcohol, mannitol, formed metabolically by the plant from fructose. Mannitol, a close cousin of sorbitol (found in apples, pears, and cherries) is used in some foods for diabetics and in certain "low sugar" gums and candies, and has very few calories. Mannitol is a very important chelating agent.

ASCOPHYLLUM nodosum supplies a wide range of carbohydrates including mannitol and organic acids. Mannitol is a known chelating agent and explains in part why seaweed can release unavailable trace elements already in the soil.

Carbohydrates	wt. % based on dry wt. of ASCOPHYLLUM nodosum
Mannitol	4.2%
Alginic Acid	26.7%
Laminarin	9.3%
Other Sugars	21.4%

GENERAL CHARACTERISTICS:

Color	Dark brown to black
Physical state	Flaky powder approximately 15% — 60 mesh approximately 99% — 15 mesh
Odor	Characteristic for processed seaweed
Density, bulk density, or specific gravity	Bulk density, approximately 500 kilos per M^3

Solubility	100% in a 50% water solution at 20'C.
pH	9.1 in a 10% water solution
Stability	Stable at normal temperatures and light conditions
Oxidizing or reducing	None
Flammability-flashpoint	None
Explodability	None
Storage stability	Very stable under warehouse conditions which reflect the normal storage conditions of the product in the original container.
Corrosion characteristics	Non-corrosive
Toxicity	Non-toxic — biodegradable
Compatibility	Compatible with most commonly used sprays

TYPICAL COMPOSITION:

Moisture Content	4%-5%
Organic Matter	50%-55%
Ash Content	45%-50%

TYPICAL MINERAL ANALYSIS:

Nitrogen	1.00%	Magnesium	0.80%	Manganese	12 ppm
Phosphorus	0.05%	Sulphur	3.70%	Zinc	100 ppm
Potassium	10.00%	Copper	5 ppm	Boron	80 ppm
Calcium	1.20%	Iron	1200 ppm		

STOCK SOLUTION PREPARATIONS:

For a 10% solution stir 12.8 oz. of **ASCOPHYLLUM nodosum** powder into 1 gallon of water. As a preservative use 0.3% Nipasol M Sodium or a similar product. Stock solution stronger than 30% total solids — not recommended.

FULL BOTANICAL NAME: **DURVILLEA** (family name) **potatorum** (species). Ref. M.N. Clayton, Monash University, Melbourne, Australia. Book title: *Seaweeds of Australia.*

Australian **DURVILLEA potatorum**
(M.N. Clayton, Seaweeds of Australia)

INTRODUCTION:

A liquid concentrate is prepared by alkaline hydrolysis of Australian Bull Kelp (Seaweed) **DURVILLEA potatorum.** Plant growth regulators are materials which are responsible and promote plant growth and plant development. Plant growth regulators are present in extremely small and balanced quantities as designed by nature. The concentration usually does not exceed parts per million (ppm) and can be as little as parts per billion (ppb).

The growth regulators (hormones) which promote growth are auxins and cytokinin like compounds.

E.G. Auxins: Tri Indole Acetic Acid

cytokinins: Trans Zeatin Riboside (ZR) 7.0 ± 1.0 (μg/litre)
Isopentenyl Adenosine (IPA) 2.0 ± 1.0 (μg/litre)
Trans Zeatin (Z) 0.7 ± 0.3 (μg/litre)
Isopentenyl Adenine (IP) 16.0 ± 1.5 (μg/litre)

Cytokinin-like compounds are responsible for cell division in plants. Tri Indole Acetic is a naturally occurring plant growth promoter for root development and bud initiation.

DURVILLEA potatorum is a suitable seaweed from the brown seaweeds to be used as raw material for the production of seaweed extract.

GROWTH PROMOTERS:

DURVILLEA potatorum contains a number of growth promoting substances in natural form.

The concentrate at 22% solids contains — 150 micrograms per litre of Tri Indole Acetic Acid

DURVILLEA potatorum GENERAL PROPERTIES:

Color	Dark brown to black
Physical State	Viscous Liquid
Odor	Typical for Processed Marine Material
pH	9.0-10.2
Chemical Stability	Stable at above pH Range and to 104° F

Below a pH level of 8.6 and between temperature of 78° F to 86° F. Biochemical activity may commence (Fermentation).

Oxidizing or reducing	None
Flammability-flashpoint	Non-flammable, no flashpoint
Explodability	None
Storage stability	Completely stable under normal conditions
Low temperature storage	Freezing commences at about 24° F in 52-gallon drums

Please ask for special report of freeze thaw stability.

Corrosion characteristics	Non-corrosive
Toxicity	Non-toxic — fully biodegradable
Chemical compatibility	Compatible with commonly used agricultural chemical compounds when mixed in recommended dilution rates with water.

Do not mix in concentrated form with any other chemical material which is also concentrated, as this may have an effect on growth regulators.

TYPICAL COMPOSITION OF 23% CONCENTRATE:

Solids Content	21%-23%
Organic Matter	16%-18%
Nitrogen (dry weight)	0.8%
Phosphorus (dry weight)	0.81%
Potassium (alginate) (dry weight)	11.4%
Calcium (wet weight)	0.24%
Magnesium (wet weight)	0.16%
Sulphur (wet weight)	0.15%
Copper (wet weight)	3.3 ppm
Iron (wet weight)	95 ppm
Manganese (wet weight)	3.5 ppm
Zinc (wet weight)	23 ppm
Boron	7 ppm

DURVILLEA potatorum concentrate has been developed and is manufactured in Tasmania, Australia.

Hans Torgersen (standing), Algea Produkter, Norway, explains types of seaweed to Dr. Adel Fouad and Dr. T. El Kobbia, Faculty of Agriculture, Ain Shams University, Cairo, Egypt; and Dr. T.L. Senn, Clemson University, Clemson, South Carolina, U.S.A.

Mr. Kristian Skou, Sales Director, Alginate Division Protan A/S Drammen, Norway, and Dr. T.L. Senn, Clemson, S.C., U.S.A., discuss **LAMINARIA** harvesting in Norway fiords.

Obtaining on-the-job information related to Norwegian **ASCO-PHYLLUM nodosum.** Harvesting is vital to Dr. T.L. Senn's expertise in seaweed usage.

CHAPTER 3
Facts to Build On

Hats off to the past; coats off to the future (Anonymous).

Utilization of seaweed as a fertilizer dates back to antiquity. With the development of chemical fertilizers in the late 1800's, this type of fertilization began to decrease in popularity. In recent years when adverse effects of the addition of many chemicals to the environment were questioned, natural sources of fertilizers or fertilizer supplements have been under consideration.

The research reported herein is for the purpose of collecting information concerning worldwide usage of seaweeds in agriculture, including field and greenhouse studies along with bioassays for plant hormones in seaweeds.

According to the literature (1, 5, 7, 9, 26, 28) seaweed extracts (either sprays or liquids) are used for the purpose of increasing the production of fruit, vegetables, potatoes, flowers, and for prolonging shelf life as well as providing crop resistance towards insects and diseases.

Senn and Kingman (30) found that solutions prepared from a dehydrated powder of Norwegian **ASCOPHYLLUM nodosum** when sprayed onto the foliage of several horticultural crops, resulted in physiological responses which exceeded those which could be explained by the chemical analysis of the seaweed. Greenhouse tests with chrysanthemum grown in varying concentrations of **ASCOPHYL-**

LUM nodosum meal resulted in significantly improved quality of chrysanthemum. Seaweed spray applications to grapes resulted in significant increases in soluble solids when compared to water sprayed controls. Field grown tomatoes and soybeans receiving weekly sprays of **ASCOPHYLLUM nodosum** extract resulted in significant increases over the control in valuable solids of tomatoes and protein content of soybean. Preliminary data of certain field-grown horticultural crops receiving foliar sprays of **ASCOPHYLLUM nodosum** indicate that this brown seaweed can play a role in a supporting capacity for reduction in commercial fertilizer application with no reduction in yield or quality of the crops. Falgout (12) found no significant differences from controls in yield and quality of sugar cane, yield of soybean, and several special crops when spray treated with a **Sargassum sp.** seaweed. The researcher, however, stated that it is quite possible that spraying earlier in the season could have caused more activity in these crops tested.

The growth promoting substances in fresh algae have been reported (2, 3, 5, 6, 20). Evidence suggests the presence of a growth hormone(s) of some type in the seaweed extracts. Jen (16) reported that liquid extracts of **ASCOPHYLLUM nodosum** contain auxins or auxin-like compounds. However, no detailed quantitative measurements were made. He concluded also that the benefit of seaweed extracts could arise from physiological changes in the internal structure of plant cells which could be induced by the high contents of the micro-nutrients in the seaweed extracts.

Other research indicates the presence of additional substances which might have significance in the plant growth promoting characteristics of seaweeds.

Material presented by Munda and Gubensek (23) includes the following: "The total amino acid pool of 10 **Rhodophyceae,** 10 **Phaeophyceae,** and 4 **Chlorophyceae,** common in Icelandic waters, was determined in the acid hydrolysates of the seaweed meal on an JEOL JLC-3 BC2 amino acid analyser."

Seventeen of the common amino acids have been determined (23). The results indicated rather pronounced differences between **Rhodophyceae, Chlorophyceae** and **Phaeophyceae,** cystine being absent in the latter group.

The distribution pattern of amino acids within different taxa of the brown algae seems more uniform than within the red algae, in spite of the pronounced differences in the anatomical structure and ecology of the species investigated.

In most of the species examined, aspartic and glutamic acid are the major constituents, and next to them alanine. The latter amino acid is predominant in two **Porphyra** species.

Variations in the cystine content were detected among the red algae, the highest amounts being present in **PLUMULARIA elegans** and **CYSTOCLONIUM purpureum.**

The average contents of the main acidic amino acids present in red algae exceed 2-3 times the average content found in the brown algae.

Among the green algae, **CLADOPHORA rupestris** deviates from the rest of the species by its high nitrogen content and exceptionally high amount of sulphur containing amino acids.

In the edible species, the essential amino acids are present in notable quantities (23).

According to Mateus, et al. (20): The amino acid composition of the protein hydrolyzate of the whole giant kelp **MACROCYSTIS pyrifera** was determined. As a percentage of the dry weight of the seaweed corrected for 12.8% moisture, the amino acids found were: alanine 2.13%, ornithine 2.11%, glutamic acid 1.98%, aspartic acid 1.45%, valine

1.06%, leucine 0.89%, glycine 0.76%, phenylalanine 0.63%, threonine 0.63%, lysine 0.60%, isoleucine 0.55%, serine 0.51%, arginine 0.36%, histidine 0.21% and methionine 0.19%. Proline and cystine were found in trace amounts.

In a study conducted by Fries (14) it was found that: Many phenylderivatives of acrylic, pyruvic, acetic acids and also derivatives of simple benzoic acids, which are naturally occurring in higher plants, influence growth and morphology of multicellular algae at hormonal concentrations, 1.10^{-5} down to 1.10^{-8}M. Indolyl-acetic acid and also phenylacetic acid (PAA) and p-hydroxyphenylacetic acid (p-OHPAA) in the brown alga **UNDARIA pinnatifida** were identified. By bacterial or fungal activities the phenylderivatives were known to be transformed into these phenylacetic acids. These two substances have been known since the fifties to give hormonal effects in many higher plants. The response of many biotests to these substances are quite the same as to IAA, but they were looked upon as artificial auxins until these phenylacetic acids were identified in a great number of higher plants only a few years ago.

The manurial value of seaweeds has long been reported. The greatest coastal applications of seaweed for manure today are along the entire coast of Brittany, and the harvesting and use of the weeds are rigidly controlled (30).

Commercial seaweed extracts became available in 1950 — the most commonly used weeds being **ASCOPHYLLUM nodosum** order FUCACEAE. Seaweed extracts are generally applied as foliar sprays. When foliar feeding became accepted as orthodox practice in the 1960's, this notable change boosted the interest in liquid seaweed products throughout the world.

In recent years, the results of scientific research provided evidence that seaweeds contain more than 70 microelements, and that their representation in these plants is considerably higher than it is in terrestrial plants. Of organic substances, marine algae contain, in addition to carbohydrates, proteins, fats and vitamin substances of a stimulating and antibiotic nature.

According to Blunden (6) results are given of field trials using a commercial seaweed extract, prepared from species of **Laminariaceae** and **FUCACEAE,** as a fertilizer additive. Foliar application to banana plants decreased the time to shooting and increased the average bunch weight of the fruits. No apparent difference between the control and test plants could be detected in the leaf content of nitrogen, phosphorus, potassium, calcium and magnesium, but marked differences were observed in the uptake of manganese. Soil application of the seaweed extract in gladioli trials increased significantly the average corm weight. Increased crop yields were obtained in commercial trials with potatoes, sweet corn, peppers, tomatoes and oranges. Improved storage quality was noted of peppers and oranges from plants treated with seaweed extract.

Reports that seaweed releases unavailable minerals from the soil have been made. Franki (13) found that leaves of tomatoes treated with seaweed meals and sprays contained more manganese than was present in the seaweed itself, concluding that seaweed had released unavailable manganese from the soil. Offermans (24) found iron availability greatest in soil to which seaweed with or without $FeSO_4$ had been applied. Results indicated also that inorganic P of seaweed origin was probably in a form other than simple phosphate. Senn et al. (28) reported that apparently concentrations of seaweed extracts at 1:10 (1 part seaweed: 10 parts water) can supply enough magnesium, manganese, zinc, and boron for pineapple orange seedlings (**CITRUS sinensis** var.) to carry on near normal respiration. Aitken (1) made similar observations. Lynn (19) studied the chelating properties of

ASCOPHYLLUM nodosum by adding the seaweed extract at concentrations of 1:50 and 1:100 to mineral deficient nutrient solutions used on pepper plants **(CAPSICUM annuum)** grown in sand. His findings, although at higher seaweed levels than those reported herein, showed improved plant utilization of boron, copper, iron, manganese, and zinc. He proposed that the improved utilization was due either to chelation or improved metabolic activity. Magnesium and calcium foliar content were not significantly affected by the seaweed.

Senn, et al. (28) reported that as the concentration of seaweed extract increases, the rate of respiratory activities of seeds also increases. Aitken (1) found that at the highest concentrations studied (1:10; 1 part seaweed; 10 parts water), seed germination was low.

Fruit harvested from peach trees which had been sprayed with seaweed extracts had increased shelf life when compared to fruit from untreated trees (29). Povolny (27) found that 0.04% solution of Norwegian dehydrated seaweed extract from **ASCOPHYLLUM nodosum** sprayed at weekly intervals onto cucumbers during fruiting increased yield 41.8%. Storage life was prolonged 14-21 days and the cucumbers were more resistant to softening and rotting than control plants. His results were obtained using both transplanted and direct-sown seed plants over 3- and 2-year periods, respectively.

Several groups of investigators have demonstrated with the aid of analytical techniques, sometimes different, the existence of cytokinins in the thallus of 10 marine algae and algae in fresh water (2, 3, 15, 17, 25, 31, 32). These substances seem to be largely distributed since they have been found with the representatives of **Phaephycophyta, Rhodophycophyta,** and **Pyrrophycophyta, Chrysophycophyta,** and the **Chlorophycophyta.** Only the representatives of the branching off of the **Cyanophycophyta** have not yet been considered. These encouraging results most normally create putting into plan new experiments conducted in the identification of cytokinins present in the thallus of algae; these experiments contribute the necessary complement to the work of Pedersen which has led to the identification of a cytokinin present in sea water bathing the algae.

Senn and Kingman (30) found that when certain species of marine algae are dried, ground, and added to media in which plants are growing, or are sprayed onto the foliage as aqueous suspensions, effects on plant growth resemble classical hormonal responses. They observed auxin, cytokinin, and gibberellin-like responses in **ASCOPHYLLUM nodosum** extracts using bioassay techniques. They used three standard bioassay systems to test for plant growth hormones in extracts of **ASCOPHYLLUM nodosum**. The **Avena** straight growth test showed that seaweed concentrations 100 μg/ml elicited auxin-like responses from coleoptile segments. The tobacco callus tissue test for cytokinins revealed significantly greater proliferation of callus than the control or the kinetin standard when treated with 10 μg/ml, 50 μg/ml, or 100 μg/ml of **ASCOPHYLLUM nodosum** extract. The measurement of gibberellin-induced a-amylase release from barley endosperm was used to measure a gibberellin-like substance in **ASCOPHYLLUM nodosum.** Significant increases in amylolytic activity over the control of the gibberellic acid (GA$_3$) standard with seaweed concentrations of 10 μg/ml, 100 μg/ml, 500 μg/ml or 1000 μg/ml indicated the presence of gibberellin-like substances in a dehydrated powder of **ASCOPHYLLUM nodosum.**

The seaweed of **ASCOPHYLLUM nodosum** utilized for agricultural purposes, contains compounds of indoleacetic acid, alginic acid and its salts. The presence of gibberellin, cytokinin and substances of a similar composition has been confirmed for algae of species of **LAMINARIA-CEAE** and **FUCACEAE** by a number of authors (4, 7, 18, 22).

According to Blunden and Wildgoose (7) foliar application of an aqueous seaweed extract of known cytokinin activity produced a significant increase in yield of potatoes of the variety King Edward. A synthetic cytokinin, kinetin, also produced significant increases in the yield of tubers of both the varieties King Edward and Pentland Dell. There was close correlation between the results from the use of kinetin and the seaweed extract of equivalent cytokinin activity, which suggests that the effect of the seaweed extract was due to its cytokinin content. In a complementary trial with King Edward potatoes, significant increases were obtained again in tuber yields through treatment with aqueous seaweed extract. The major increase was in potatoes.

Blunden and Nicholson (8) presented results of field trials on varieties of sugar beet using a commercial aqueous seaweed extract prepared from species of **LAMINARIA-CEAE** and **FUCACEAE**. In the seaweed-extract-treated plants, significant increases were obtained in root weight, root sugar content and in clarified juice purity. In addition, the amino-nitrogen and potassium contents of the extracted juice were significantly reduced. Significant results were obtained when the extract was applied in late May, but no significant effect was recorded when the extract was applied in late April. The results which were obtained are discussed in relation to the known high cytokinetic activity of the seaweed extract.

According to Williams, et al. (32) the three commercial seaweed extracts most widely used in the U.S. have been tested for the presence of auxin, gibberellin and cytokinin activities. No significant auxin activity was detected in any of the extracts. Kingman and Senn (18) observed that upon oxidation, activity if present could have been lost. Gibberellin-like activity was demonstrated in freshly manufactured samples of all three commercial products. However, this activity declined rapidly, and within four months was negligible. Highly significant cytokinetic activities were found in all three extracts although large variations were found between different batches of the same product. The gibberellins are less stable than cytokinins; therefore, time could have contributed to further hormonal destruction.

Following the memorable experiences of Went and with numerous effective works for 20 years, we actually know that tests for auxins and the development of superior plants are not only conditioned by the quantity and quality of the nutritive matters absorbed by the roots or synthesized by the leaves, but also by endogenous substances.

Augier (2, 3) indicated that the action of phycological extracts on algae in culture and on other plants, just like the action of extracts of diverse plants on algae, shows that growing thallophytes must be under the dependence of substances of hormonal nature. This hypothesis is reinforced by studies carried on tropisms, periodisms, regeneration and apical control.

The first assay tests of verification of this theory have been evidenced through the wheat coleoptile test. Most recent analyses have revealed the existence of numerous auxins in the thallus of algae belonging to the various taxonomic, antomic, physiological and ecological unity and to different periods of their development. These experiences have permitted identification, more or less with certitude, of the acid phenyl acetate and the acid indole-3-glycolic as well as various other indole compounds. Studies have also revealed other substances equally very active on growing plants but whose chemical composition remains unknown.

Finally, the distribution of auxins in the thallus have been well studied with some species just as the modality of their anabolism and their catabolism.

Augier (2, 3) further explored attempts at identification of gibberellins, cytokinins, and various other substances of hormonal nature.

Mowat-Bentley (21) and Mowat and Reid (22) demonstrated the existence of substances similar to cytokinins with unicellular marine algae **GYMNODINIUM splendens** and **PHAEODACTYLUM tricornutum** as well as on a sample of marine phytoplankton including especially the diatoms. The extraction includes essentially treatment with selective solvents and chromatography. The algae once extracted in aqueous acetone (90%) are filtered and the filtrate dissolved in benzene in successive stages. The extract is then chromatographed on paper and the chromatograms are then analyzed by the biological test, leaf of radish, a test which is unfortunately not entirely specific to cytokinins. The preparation by the biological test equivalent to kinetin gives concentrations which vary from 0.1 to 10 micrograms per kilogram of fresh algae.

Augier (2, 3) states that numerous laboratory experiments with cultures, and some tests in the sea have shown that auxins, gibberellins, cytokinins and several substances of hormonal nature play a part in the regulation of the great physiological functions of these algae during the different stages of their life. Thus, phytohormones participate in the regulation of growth and development affecting not only the mechanisms of cellular elongation and division but also those of morphogensis, organogensis and cellular differentiation. They also play a part in other physiological mechanisms of development, e.g., tropisms, periodisms, regeneration and apical dominance. Actually little is known about the precise modalities of plant hormone action, the place, and their exact importance in the different mechanisms of cellular elongation, division and differentiation in algae.

Pedersen (25) detected a cytokinin in sea water collected from the **Fucus-ASCOPHYLLUM** zone. Cytokinins have been extracted from a one-celled alga **(LAMINARIA spp.)** (22), from **LAMINARIA digitata** (15), and from **Ectonia sp.** and **Hyphnae sp.** (17).

Brian, et al. (10) reported that cytokinin activity of a commercial aqueous seaweed extract was demonstrated by the promotion of growth **in vitro** of carrot explants in a cytokinin-free medium containing the seaweed extract. Cytokinin activity of the seaweed extract was confirmed by the increase in growth it produced with tissue-cultured cells of a cytokinin-requiring strain of **ATROPA belladonna,** and also by the radish leaf bioassay. The results showed that the seaweed extract has a cytokinin activity capable of producing physiological changes, even when applied at the low concentrations used in practice.

Augier (2, 3) showed evidence of cytokinins in the thallus of algae with the aid of a different method. The algae reduced in powder are treated in methanol and the methanolic extract chromatographed on the column of Sephadex gel LH 20. The elution in ethanol to 35% permits obtaining 3 to 7 fractions which are chromatographed in a thin bed of cellulose. The cytokinins are marked in the fraction of their natural fluorescence in ultra-violet and with the aid of biological tests "marrow of tobacco" and "leaf of oaks," specific to cytokinins. The most active fractions which deposited a column of active carbon are specifically chosen with the aid of methane, diethylanin and of benzene. The cytokinins contained in each elution fraction are precipitated with the aid of a solution of silver nitrate which is then eliminated by a solution of chlorhydric acid. The attempts at identification consist of comparing the Rf of the detected substances with those of the reference cytokinins, several systems of solvent as well as noting their coloration in Woods reagent. The joint utilization of these different techniques demonstrates evidence of substances of cytokinin type with the **Chlorophyta Enteromorpha linza,** the **Pheophyta Cystoseira discors** and the **Rhodophyta Gymnogongrus norvegicus.** The attempts at identification show that the substances could not be assimilated in either zeatin or in 6-benzyladenine, but they very probably carry a new purine or pyrimidine.

The presence of cytokinins in marine algae is confirmed by Brian, et al. (10) with the aid of biological tests "explants of carrot," "tissue of **ATROPA belladonna**" and "leaf of radish." The biological tests are applied directly to a sample of conversional product as the base of **Fucales** and **Laminariales.**

Hussain and Boney (15) found substances similar to kinetin in the stem and roots of the **LAMINARIA digitata.** The crude extract in aqueous methanol (70%) reduced by evaporation of methanol in the only aqueous phase is purified in petro ether then adjusted to pH 5.5. This aqueous extract is then chromatographed in a column of resin DOWEX 50 W with the aid of an ammonia eluant and the fractions are chromatographed on paper. The chromatograms are analyzed by the biological test "senescence of leaves of radish." The eluants of active zones of the chromatograms give, on the spectrophotometer, peaks of absorption at 265 nm and two summits from the roots and only one summit for the stems.

Blunden (5) states that the need for supplying seaweed extracts of known cytokinin activity is recognized and that a reliable and simple procedure for the routine assay of the extracts is a major problem. The extracts contain substantial amounts of growth inhibitory materials which interfere with many of the cytokinin-bioassays. It is possible to remove many of the inhibitors by extracting with organic solvents, and the purified material can then be successfully assayed by a variety of different methods. However, the results obtained after purification may not be applicable to the unpurified extracts which are used in agriculture. It is possible that a more reliable result is obtained by assaying the unpurified extracts, but the applicability of results obtained from biological assays carried out in the laboratory to actual effects produced in the field is uncertain. The tissue culture bioassays, such as the tobacco pith callus, tobacco stem pith, soybean callus and carrot root tissue, are at present the most reliable methods for testing crude extracts. However, these procedures utilize sophisticated techniques and are very lengthy, which reduce their suitability for the routine analysis of seaweed extracts by small manufacturers. The radish leaf expansion bioassay, although somewhat insensitive and prone to giving erratic results, has been found useful for the routine analysis of the unpurified seaweed extracts. Nevertheless, the development of a more suitable assay method which is rapid and reliable is necessary.

With the aid of the test "marrow of tobacco" Pedersen (25) showed evidence in 1972 of a cytokinin-type activity in an extract of ethyl acetate in seawater retained in the zone of **FUCUS** and **ASCOPHYLLUM.** He further succeeded in extracting a sufficient amount of active substance under the test "marrow of tobacco" to be able to identify it in the cytokinin 6-(3 methyl-2burenyllamino) purine with the aid of the chromatograph in gaseous phase and the spectrophotometer in mass.

The technique of extraction utilized by Jennings (17) was entirely different from that employed by Mowat (21). The treatment began by the extraction in ethanol, the extract being then filtered and dissolved in water at pH 3 and in ethyl acetate. The aqueous phase is deposited on a column of Amberlite resin of which the elution in NH40H HC1 gives different fractions which are then chromatographed on paper. The chromatograms are analyzed by the Wood blue reagent and the biological test senescence of the leaf of barley based on the retention of the chlorophyll by the kinetin. In this way, Jennings (17) showed evidence of a weak activity of cytokinin with **Rhodophyton hyphnae musciformis** while the extracts of **Phaephyta ectonia radiata** includes (contains) two substances similar to cytokinin and a substance inhibited by the biological test.

Seaweed research by Darrah and Hall (11) investigated the effect of seaweed applications upon the incidence of **Fusarium roseum tricinctum** and nematode populations in Kentucky Bluegrass turf. They found the following:

1. After three continuous years of 250 lb/acre applications of granular seaweed **(ASCOPHYLLUM nodosum)** on Kentucky bluegrass-perennial ryegrass fairway turf, the incidence of **F. roseum** is similar to that found in plots treated three years with Benomyl + Nemacur treated plots and 48% less than that found in control plots. Combination treatments of liquid and granular seaweed or liquid seaweed alone did not produce this effect.

2. The three-year granular seaweed treatment significantly reduced (98% reduction) the population of **Paratylenchus** spp. pin nematodes when compared with control plot counts. After three years of treatment, control of this nematode was as good as that provided by commercially available nematicides. A 38% reduction in **Pratylenchus penetrans** lesion nematode count was brought about by the granular seaweed treatment. This reduction was not statistically significantly different from the control treatment.

3. On June 27, 1973, and August 8, 1973, significant correlations were observed between **Pratylenchus pene-**

trans and the incidence of **F. tricinctum**. On June 27, 1973, and September 19, 1975, significant correlations were observed between **Paratylenchus** spp and the incidence of **F. tricinctum**. On September 12, 1974, and August 4, 1975, significant correlations were observed between **Tylenchorhynchus maxims** and the incidence of **F. tricinctum**. The correlation of both **Paratylenchus** and **Pratylenchus** nematodes with the incidence of **F. tricinctum** and the concomitant reduction of these pathogens with granular seaweed suggest the possibility that granular seaweed applications might have a long-term role to play in the reduction of the incidence of **F. tricinctum.**

4. No significant effect of seaweed applications was observed on Kentucky bluegrass foliar analysis, turfgrass quality rating, Japanese beetle grub populations, incidence of dollarspot **(Sclerotinia homeocarpa)** or stripe smut **(Ustilago striiformis),** soil pH, soil soluble salt concentration, soil organic matter, or soil nutrient availability.

Darrah and Hall (11) conducted experiments for a period of one year to study fertilizer-efficiency seaweed incorporation and nitrogen-source seaweed **Fusarium**-level relationships.

Upper, et al. (1970) suggested that since cytokinins, gibberellins, ABA, and IAA are all measurable by GLC, it should be possible to measure the entire content of known plant growth regulatory substances in various natural sources by simple, rapid techniques.

More recent (1980's) investigations by Young in Australia and Kingman and Moore at Clemson University demonstrate the applicability of the extraction and GLC methods to the extraction, purification, and quantitation of cytokinins, purine basis and their ribosides, and other plant hormones, such as IAA and ABA.

MacLeod and co-workers (1985) in Australia published

"Detection of Cytokinins in a Seaweed Extract" in Phytochemistry, Vol. 24, No. 11.

Dr. T.L. Senn discusses potentials of seaweed with the Minister of State, Pakistan; Egyptian Government Agricultural Officials; and Norwegian Government Officials.

Literature Cited

1. Aitken, J.B. and T.L. Senn. 1965. The effect of seaweed extract and humic acids on O_2 uptake of **CITRUS sinensis** seedlings grown in nutrient-element-deficient cultures. Botanica Marina. 8:144-148.

2. Augier, H. 1976. Les hormones des algues. Etat actual des connaissances. I — Recherche et tentatives d'identification des auxines. Botanica Marina. 19:127-143.

3. _____ 1976. Les hormones des alques. Etat actual des conniasances. II — Recherche et tentatives d'identification des gibberellines, des cytokinines, etat de diverses autres substances de nature hormonale. Botanica Marina. 19:245-254.

4. Bentley, J.A. 1958. Role of plant hormones in Algae metabolism and ecology. Nature. 181:1499-1502.

5. Blunden, G. 1977. Cytokinin activity of seaweed extract. Marine Nat. Prod. Chem. 337-344.

6. _____ 1973. The effects of aqueous seaweed extract as a fertilizer additive. Proc. Seventh International Seaweed Symposium.
7. _____ and P.B. Wildgoose. 1976. The effects of aqueous seaweed extract and kinetin on potato yields. J. Sci. Fd. Agric. 28:121-125.
8. _____ and F.E. Nicholson. 1974. The effects of aqueous seaweed extract on sugar beet. Proc. Eighth International Seaweed Symposium.
9. Booth, E. 1965. Frost resistance and insect pests: Seaweed has a two-way benefit. The Grower. 6:20-24.
10. Brian, K.R., M.C. Chalopin, T.D. Turner, G. Blunden, and P.B. Wildgoose. 1973. Cytokinin activity of commercial aqueous seaweed extract. Plant Science Letters 1:241-245.
11. Darrah, C.H., and J.R. Hall. 1976. Twin shields **Fusarium** nematode study. Report to Econ. Dev. Admin., U.S. Dept. of Commerce.
12. Falgout, R.N. 1977. Investigations into the effects of certain seaweed foliar spray on selected Louisiana Agricultural crops. Econ. Dev. Admin., U.S. Dept. of Com. Report. Wash., D.C.
13. Franki, R.I.B. 1960. Studies of manurial values of seaweeds. Chem. Abstr. 54:18849 d-e.
14. Fries, Lisbeth. 1976. Phenylacetic acids, native growth regulators in seaweed? Institute of Physiol. Bot. Uppsala, Sweden.
15. Hussain, A., and A.D. Boney. 1969. Isolation of a kinin-like substance from **LAMINARIA digitata.** Nature 223:504-505.
16. Jen, J.J. 1972. The effects of seaweed on plant growth. S.C. Agr. Expt. Sta. Hort. Res. Series No. 141. Clemson Univ., Clemson, S.C.
17. Jennings, R.C. 1969. Cytokinins as endogenous growth regulators in the algae **ECTONIA (Phaephyta)** and **HYPHNAE (Rhodophyta).** Aust. J. Biol. Science 22:621-627.

18. Kingman, A.R. and T.L. Senn. 1977. Bioassay systems to test for plant growth hormones in extracts of **ASCOPHYLLUM nodosum.** Ninth International Seaweed Symposium. Santa Barbara, Calif.

19. Lynn, L.B. 1972. The chelating properties of seaweed extract **ASCOPHYLLUM nodosum** vs. **MACROCYSTIS perifera** on the mineral nutrition of sweet peppers **CAPSICUM annuum.** M.S. Thesis, Clemson Univ., Clemson, S.C.

20. Mateus, H., J.M. Regenstein, and R.C. Baker. 1975. The amino acid composition of the marine brown alga **MACROCYSTIS pyrifera** from Baja, California, Mexico. Botanica Marina, 19:155-159.

21. Mowat, J.A. 1964. Auxins and gibberellins in marine algae. Botanica Marina 8:149-155.

22. _____ and S.M. Reid. 1967. Investigations of radish leaf bioassay for kinetins and demonstration of kinetin-like substances in algae. Ann. Rev. Bot. 32:23-32.

23. Munda, M. and F. Gubensek. 1975. The amino acid composition of some common marine algae from Iceland. Botanica Marina 19:85-92.

24. Offermans, C.N. 1968. Effect of brown algae **(MACROCYSTIS integrifolia)** in increasing iron availability of a calcareous soil. Chem. Abstr. 68:104216.

25. Pedersen, M. 1973. Identification of a cytokinin, 6-(3 methyl-2 butenylamino) purine in sea water and the effect of cytokinins on brown alga. Physiol. Plant. 28:101-105.

26. Povolny, M. 1969. Investigations on the effectiveness of seaweed extract on yield and quality of pickling cucumbers. Hort. Abstr. 64:857.

27. _____ 1969. The effect of seaweed extract on the harvest, ripening and shelf life on tomatoes. Department of Horticulture Publ., Agricultural University, Prague, Czechoslovakia.

28. Senn, T.L., B.J. Skelton, and J.A. Martin. 1960. The effect of kelp meal on development and composition of various vegetable and special crops. Proc. Assoc. Sou. Ag. Wkrs. 57:182.

29. _____, and B.J. Skelton. 1968. The effect of Norwegian seaweed **(ASCOPHYLLUM nodosum)** on the development and composition of certain horticultural crops. S.C. Agri. Expt. Sta. Hort. Res. Series No. 23. Clemson Univ., Clemson, S.C.
30. _____, and A.R. Kingman. 1977. Physiological responses of certain horticultural crops to applications of meal and extracts, **ASCOPHYLLUM nodosum.** Ninth International Seaweed Symposium. Santa Barbara, California.
31. Sivalingam, P.M., T. Ikawa and K. Nisizawa. 1976. Physiological roles of a substance 334 in algae. Botanica Marina 19:9-21.
32. Williams, D.C., K.R. Brian, G. Blunden, P.B. Wildgoose, and K. Jewers. 1974. Plant growth regulatory substances in commercial seaweed extracts. Eighth International Seaweed Symposium. Bangor, Wales.

Audience of scientists, faculty, government, and growers attend seaweed conference at NARC in Karachi and Islamabad, Pakistan.

CHAPTER 4
What Makes Plants Grow?

Many things grow in the garden that were never sowed there (Proverb).

All living things which contain chlorophyll (the green coloring matter in plants) have the ability to transform the kinetic energy of light into the potential chemical energy in foods and other manufactured compounds. Man has always had a serious problem in obtaining an adequate food supply. He has continually tried to improve the efficiency of plants that play such a vital role in his life-style. The key to solving the world's food supply is understanding the mechanisms that regulate the metabolism of plants. These changes in cells and organs provide energy for growth, development, and reproduction in plants. As man develops a more complete understanding of factors affecting growth processes in plants, especially on hormonal control, he will be able to answer the question "What makes plants grow?"

The plant is the ultimate source of all of the world's food; even our drinks — coffee, tea, cocoa, lemonade, and in a round-about way, milk — which comes from plant-eating animals. Most of our clothing — cotton, linen, rubber — comes from plants; and wool and silk come from plant-eating animals.

Plant Processes

The study of the processes by which these substances, which serve man in so many ways, are built up from the elements is exceedingly interesting and highly important. Apart from the utility of the study there is also an aesthetic value.

We are slowly learning that we may guide some of the forces, which have been throughout all the ages, performing this beneficent work which we see everywhere — above, beneath, and around us — in the vast teeming plant life of the world.

From early times, people have been interested in the movements of plants. The turning of stems toward the light, the drooping of the leaves of the sensitive plant on contact, the movements of tendrils and twining plants — all these have attracted the attention of the observers for many years. These movements of plants are all brought about by growth regulators, some of the same growth regulators that occur in **ASCOPHYLLUM nodosum** — an important brown seaweed.

Plant Growth

Researchers are interested in and are investigating the chemical and physical changes within the organs which bring about plant growth. They must take advantage of all the methods and apparatus available from the science of chemistry, physics, nuclear engineering, computer science, and all branches of agricultural science to obtain the answers to "What makes plants grow?"

Growth is a subject of interest to all of us. It is a personal subject. We have all grown. "What is growth?" Most living things originate as a single cell, microscopic in size and derived from the parents.

One of the most interesting and hopeful things we have learned is that growth is fundamentally the same process in humble and simple living things, both plants and animals, as it is in man. It is important, therefore, to study growth in plants not only because such study enables us to control these living things and use them, but also because it tells what may occur in our own bodies.

Have you ever asked, "Exactly what is a plant and what makes it grow?" The answer is not so simple when one studies the plant life from seed to flower to fruit and then senescence and death.

Seed

The most highly developed group of plants are called seed plants. This group contains more members than all other plant groups put together. Seeds vary in size, shape, and color. They are all alive and capable of growth if given favorable conditions under which to germinate. These factors are moisture, temperature, and sometimes light. All aid in the activation of plant growth hormones which trigger growth. Seed treated with seaweed extracts respire much faster and will germinate sooner than non-treated seed.

Seaweed extract increases seed activity. It has been previously reported by Booth that seaweed products might hasten the germination of seeds, an effect which may be attributed to the auxin or growth hormone content of the seaweed.

Based on these reports, research was initiated at Clemson in 1960 to investigate the effects of various extracts on the subsequent respiratory activity of seed and their ultimate germination potentials. The effect of seaweed extracts on the germination of beet **(Beta vulgaris)** seed is very evident. At the end of one week the germination percent of beet seed treated with seaweed extract was 84 as compared to 0 for the checks. Another experiment was conducted at the same time to study the effects of soaking beet seed in seaweed extracts for 30 minutes, prior to germination. Prior soaking of seeds for 30 minutes increased the germination of beet seed by 25% over the control. Fifteen mls of extract were used in each petri dish of 50 seed. These treatments were replicated throughout the experiment.

To further test the effect of seaweed extracts on seed activity, an experiment was initiated to manometrically measure seed response. The manometric apparatus equipped with constant volume Warburg respirometers used was the refrigerated-heated-shaker type. All tests were made at 30° C and oxygen uptake and carbon dioxide evolution were recorded. In this report only carbon dioxide evolution is reported. The measurements were made both on fresh and dry weight basis but are reported as cu. mm. of carbon dioxide evolved per gram per hour on a dry weight basis unless otherwise stated.

Treatment of several species of seed greatly accelerated the respiratory activity of the seed. The germination percentage of the treated seed was also increased. Five milliliters of seaweed extract was applied to the seed in a petri dish in varying concentrations. The concentrations varied from the pure extract through 1 to 5, 1 to 10, 1 to 25, 1 to 50, 1 to 100, 1 to 200, 1 to 300, 1 to 400, and 1 part seaweed extract to 500 parts of water.

The seed treated with the seaweed extracts were zinnia **(Zinnia elegans),** tobacco **(Nicotiana tobacco),** peas **(Pisum sativum),** turnips **(Brassica rape),** tomato **(Lycopersicon esculentum),** radish **(Raphanus sativus),** cotton **(Gossypium herbaceum),** white pine **(Pinus alba),** loblolly pine **(Pinus taeda),** Ligustrum

(Ligustrum lucidum), nandina **(Nandina domesticum),** and American holly **(Ilex Opaca Hume** No. 2).

In all cases tested, an application of the extract, even in low concentrations, increased the respiratory activity of the seed. The higher the concentration of extract, the higher the rate of respiratory activity. In the higher concentrations, where the respiratory activity was high, the germination percentage was low. However, when the concentration was such that the respiratory activity was only moderately increased, the germination percentage was increased.

The optimum concentration varied considerably with different species of seed. The optimum was generally at a concentration between 1 to 25 and 1 part extract to 50 parts of water.

A seaweed extract appears to contain potentials as a promoter of germination of certain seed (Tables 1 and 2).

The effects of seaweed extract on the subsequent respiratory activity of loblolly pine seed are shown in Tables 3 and 4. The results of a one-hour run are shown in Table 3. The run was continued for a six-hour period and the results are presented in Table 4. With increases in concentration of seaweed extract there was a resulting increase in respiratory activity. At the end of the six-hour run the higher concentration treatments were still causing increased respiratory activity.

Table 1
The effect of varying concentrations of seaweed extract on the subsequent respiratory activity of **Ligustrum lucidum** seed.
January 2, 1961

Carbon Dioxide Evolution

		Cu.mm./gram/hour			
		1-100 Extract	Check	1-25 Extract	1-5 Extract	Pure Extract
Treatment	cu.mm.CO_2 /gram/hour	37.2	42.9	62.4	114.7	164.6
		Difference between means				
1-100 Extract	37.2	—	5.7	25.2*	77.5**	127.6**
Check	42.9		—	19.5*	71.8**	121.7**
1-25 Extract	62.4			—	52.3**	102.2**
1-5 Extract	114.7				—	49.9**
Pure Extract	164.6					—

* and ** indicate significance at the 5% and 1% level, respectively.

Table 2
The effect of varying concentrations of seaweed extract on the subsequent respiratory activity of **Nandina domesticum** seed.
January 2, 1961

Carbon Dioxide Evolution

		Cu.mm./gram/hour				
		1-100 Extract	Check	1-25 Extract	1-5 Extract	Pure Extract
Treatment	cu.mm.CO_2 /gram/hour	22.8	24.3	41.3	218.8	601.4
		Difference between means				
1-100 Extract	22.8	—	5.5	18.5*	196.0*	579.6**
Check	24.3		—	17.0*	194.5**	577.1**
1-25 Extract	41.3			—	177.5**	560.1**
1-5 Extract	218.8				—	382.6**
Pure Extract	601.4					—

* and ** indicate significance at the 5% and 1% level, respectively.

Table 3
The effect of seaweed extract on the subsequent respiratory activity of Loblolly pine seed.
September 30, 1960

Treatment	Cu.mm. CO_2 evolved per gr per hour	Check	1-100	1-25	1-5
		Cu.mm. CO_2 evolved per gr per hr.			
		17.3	32.2	43.1	61.7
		Difference between means			
Check	17.3	—	14.9*	25.8**	44.4**
1-100	32.2		—	10.9	29.5**
1-25	43.1			—	18.6*
1-5	61.7				—

* and ** indicate significances at the 5% and 1% level, respectively.

Table 4
The effect of seaweed extract on the subsequent respiratory activity of Loblolly pine seed.
September 30, 1960

Treatment	Cu.mm. CO_2 evolved per gr per hour	Check	1-100	1-25	1-5
		Cu.mm. CO_2 evolved per gr per hr.			
		103.5	193.1	258.5	369.9
		Difference between means			
Check	103.5	—	89.6*	155.0*	266.4**
1-100	193.1		—	65.4	176.8**
1-25	258.5			—	111.4*
1-5	369.9				—

* and ** indicate significances at the 5% and 1% level, respectively.

The Use of Norwegian ASCOPHYLLUM nodosum for Pregermination of Onion Seeds

Cook, R.L. 1982. Mich. State Univ., East Lansing, Mich. Hort. 457

Abstract

The use of Norwegian **ASCOPHYLLUM nodosum** seaweed extract was evaluated as a pregerminating medium for onion seeds. This was carried out at 5°, 10°, and 20° C using three different concentrations of extract plus a control. Evaluations were made for: days to 50% germination, spread of germination in days (10%-90%), radicle length when growth ceased due to moisture stress, and total percent germination. Treatment at 10° C demonstrated the greatest variation from the control and positive correlation was seen between concentration and germination rate in several other samples at the other temperatures. Norwegian **ASCOPHYLLUM nodosum**, though showing in these tests to be somewhat temperature specific, does give an increase in germination as well as a consistent radicle length.

Root

When a seed sprouts, the first thing to break through is the root. No matter what the position of the seed when it is planted, whether sideways or upside-down, the root turns downward. This is due to the influence of growth regulators.

Very important parts of the root are its tip and the root hairs just behind the tip. One seldom sees them, for when a plant is pulled up, they are broken off. These parts are continually wearing off as the root penetrates the soil and therefore must be replaced. Anchorage, absorption, and storage are the main functions of plant roots. The root hairs take in enormous quantities of water and nutrients that are passed into conducting vessels for transfer to above ground parts. It is very important that micronutrients (manganese, zinc, copper, iron) be available to the plant at this stage of development. Various nutrient elements play a vital role in cell development and enlargement.

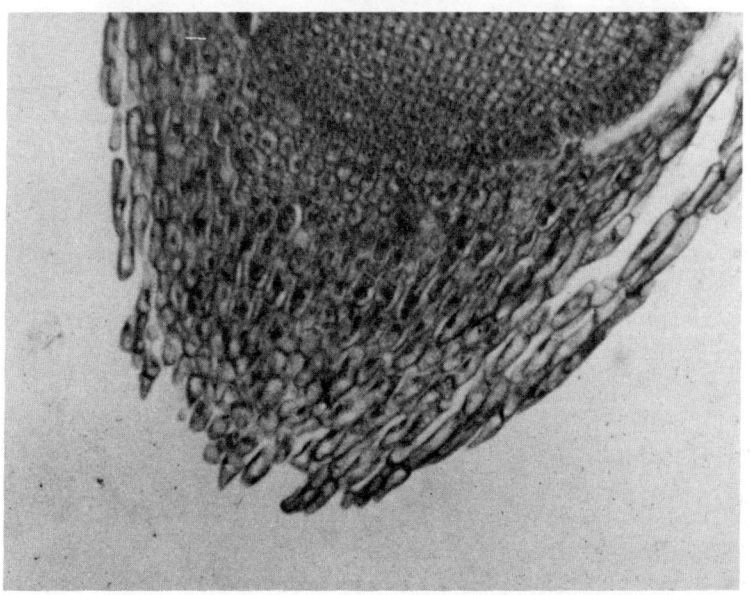

Root tip section showing wearing away of cells as root penetrates soil. Cell division must take place rapidly. Therefore nutrients and growth-promoting substances are needed at this stage of growth.

The effects of substances of a cytokinin nature (seaweed) on admission of nutrient elements into the roots of plants has been the subject of many recent research projects. It has been shown that cytokinins can have a direct bearing on admission of nutrient elements into the roots of plants. These research results are usually presented in a highly technical way which is beyond the scope and purpose of this publication. It may be simply pointed out that concentrations of cytokinins in the roots of plants accelerates admission of ni-

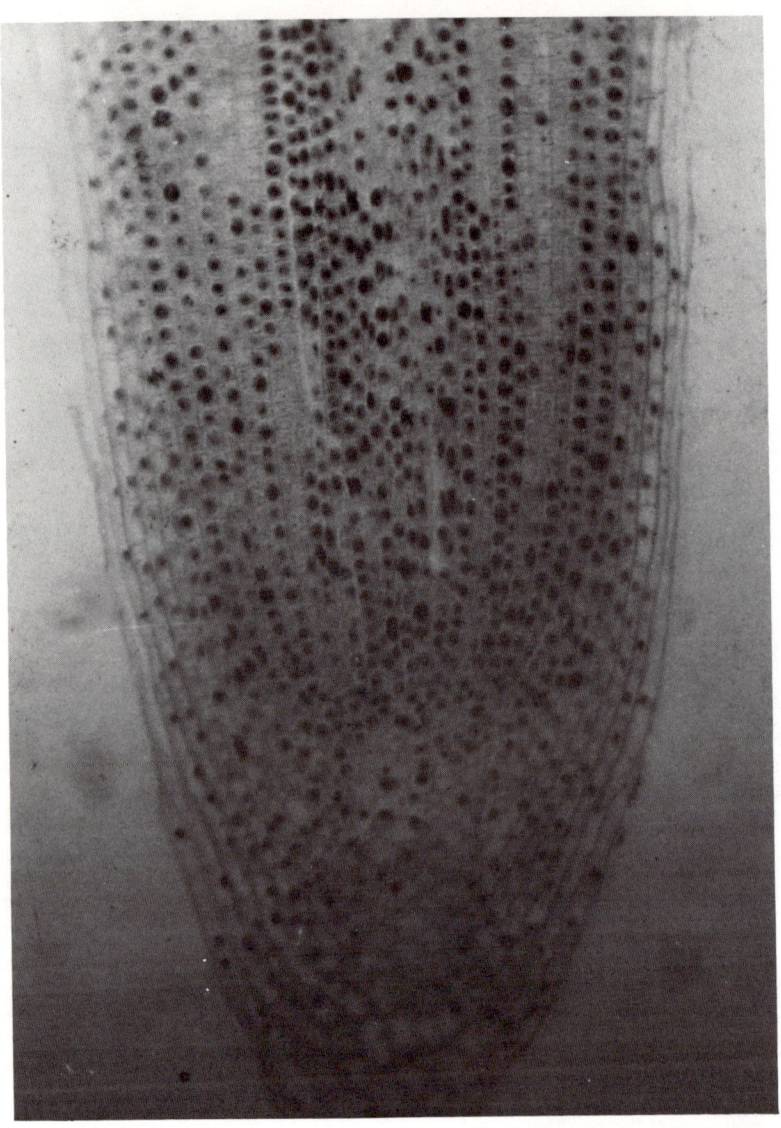

Root section showing region of rapid cell division and cell elongations. Periods of stress slow down root growth and in turn slows down stem, leaf, and flower growth. Roots growing in soil that has received applications of **ASCOPHYLLUM nodosum** will withstand stress and grow at a rapid rate.

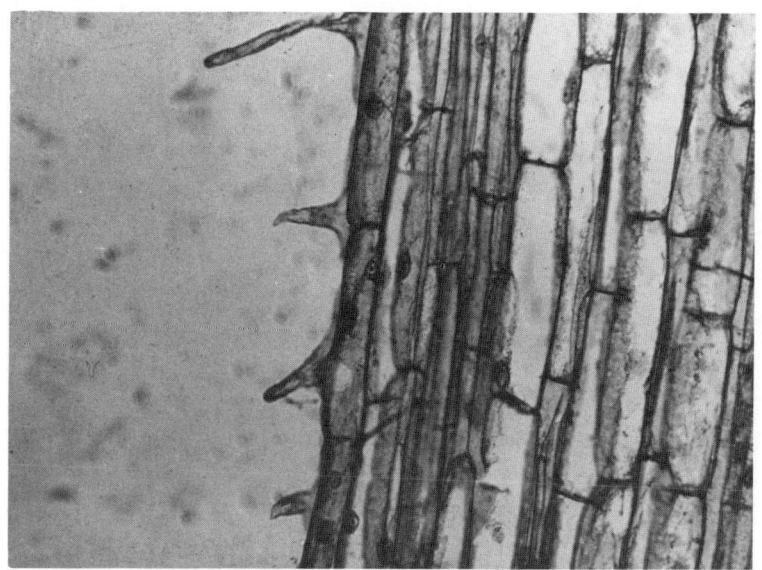

Root section showing root hairs that take in water and nutrients which enter conducting vessels that translocate them to leaves and growing areas. Cytokinins have a direct bearing on admission of nutrient elements into the roots of plants.

trates, phosphorus, potassium and calcium into the plant roots. Thus, by improving the supply of building and energy materials available to the roots, cytokinins can be conducive to absorption and utilization of nutrient elements by the plant. (See diagram.)

Stem

This brings us to another part of the plant, its transportation system. Inside the stems are tubes, called vascular bundles. The arrangement of the vascular bundles varies with the different plants. The dicots (beans, most trees and many flowers) have bundles arranged in a circle around a central pith. The monocots (corn, lily, palms) have the bundles in the form of strands running the length of the stem. The strength of the stem is very important and early applications of seaweed extract have resulted in stronger stems that are able to withstand the stress of wind and rain. (See diagram.)

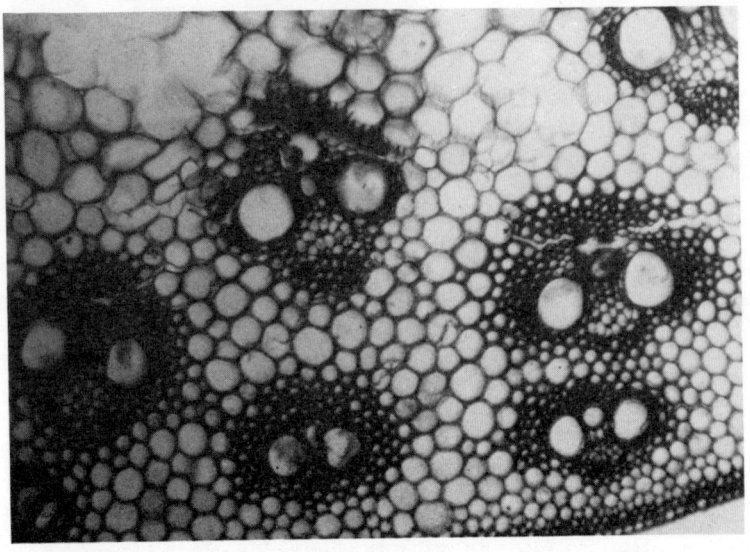

Cross section of a plant stem showing conducting vessels (vascular bundles) that translocate water and nutrients to above ground plant parts. Cell wall thickness and flexibility are very vital to plant growth during periods of plant stress.

Leaf

The structural framework of all plant material is called microfibrils. They are long interwoven and interconnecting cellulose strands made up of many long starch threads. Growth hormones are responsible for getting the microfibril material to each new plant cell. The strands are held together by calcium pectate. This gives the plant strength and yet permits the plant to bend with the wind without breaking. Thus, the nutrient elements calcium and potassium are very important to the plant in the growth process. Hormones that are stimulated by seaweed also play a vital role in how plants grow.

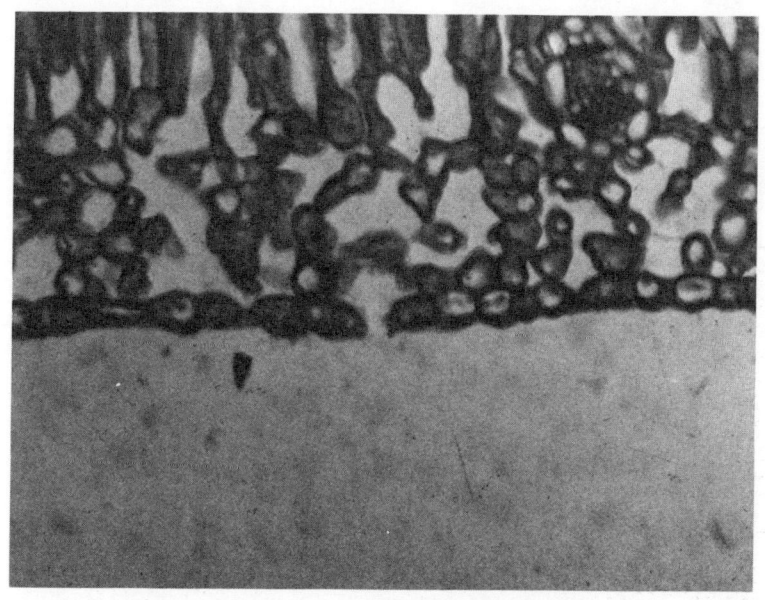

Cross section of leaf showing stomata (openings) that permits exchange of gases. These openings close during periods of stress. Manufacturing cells (palisades) are seen at the top of the photo.

Everyone has noticed that the upper surface of a leaf is usually greener than the lower surface. You will notice too that leaves turn their upper surface toward the sun. The reason is that most of the green material, called chloroplast, is located in manufacturing cells in this area. The leaf is not solid but contains many small openings that permit transfer of gases and water vapor. Elements may also enter the leaves through the cell walls. (See diagram.)

In order to make cellulose, to build new cells, to store reserve food, to carry on all life processes, a plant needs energy. Energy is obtained by oxidation of the sugars formed from photosynthesis. This oxidation process is called respiration. These are all enzymatic processes. Micronutrients, hormones, auxins, and growth regulators are all very necessary at this stage of growth. Seaweed extracts contain these vital substances and when applied to plants improve growth and development.

Plants absorb nutrients through their leaf and stem tissues as well as by root intake. It is common practice by growers to apply micronutrients by foliar applications to correct micronutrient deficiencies. The time of application is very important. The introduction of chelates has improved the success in correcting this type of deficiency.

Seaweed has its own chelating agent, mannitol, a simple sugar. Research findings have substantiated this postulation that was made in Research Series #23 from Clemson University.

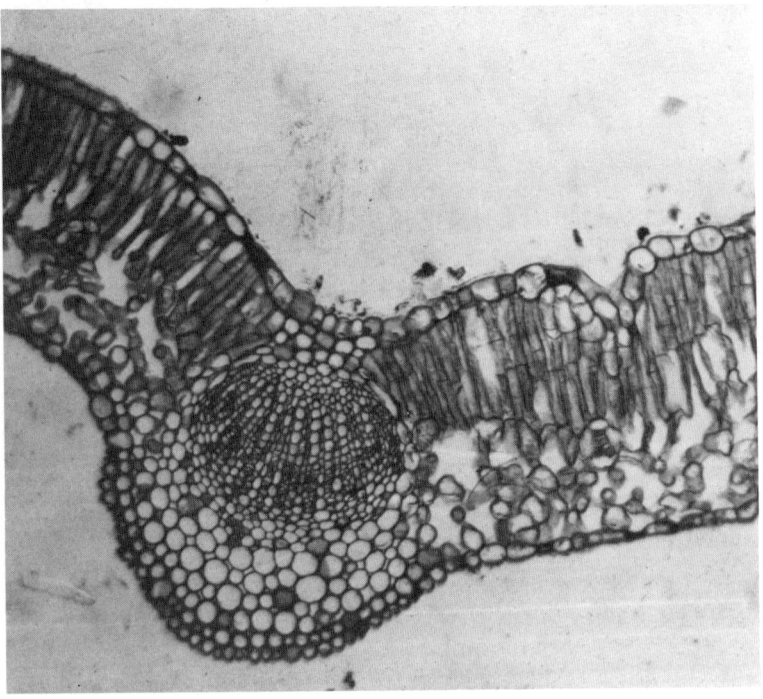

Cross section of leaf showing conducting vessels in mid-rib. Long cells arranged in rows are cells that manufacture sugar from air, water and nutrients, with the aid of sunlight energy. Stomates (openings) permit gas exchange.

Flower

We admire the bright colors and beautiful forms of flowers and take great pleasure in them. However, the flowers serve a very vital part in the perpetuation of plant life. When you look into a flower you see the various sexual mechanisms. The arrangement of these sexual organisms varies with the different plants.

The stamen contains the male organs which shed pollen grains. These grains germinate on the stigma of the pistil and grow down the style to fertilize the eggs in the ovary. This results in the formation of fruits and seeds. (See diagram.)

Female (divided five segment stigma) pistil at top and style (stalk) containing male (stamen) pollen. Pollen germinates producing tube that grows down style to fertilize eggs in embryo.

Pollen germination is stimulated by certain nutrient substances, these include manganese sulfate, zinc sulfate, calcium, iron, boron and some organic substances. These sub-

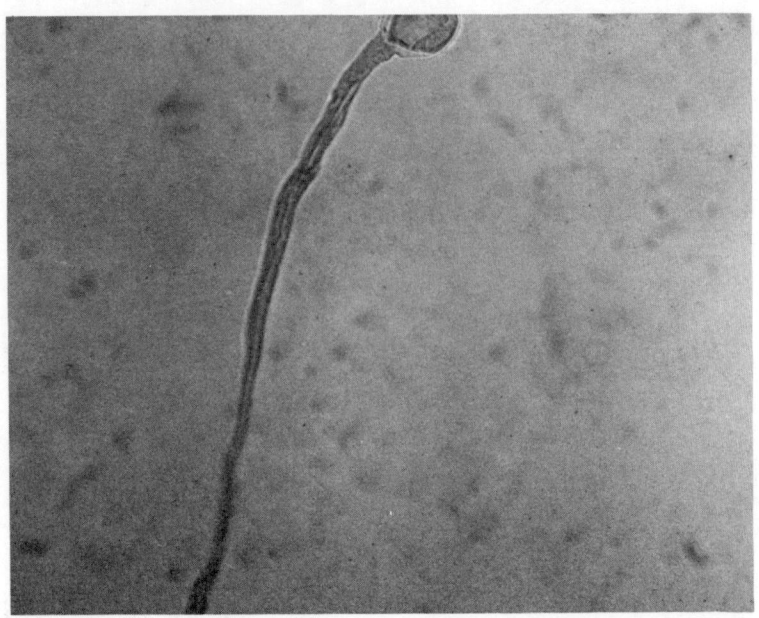
Pollen grain that has germinated producing tube that grows down style to fertilize egg in embryo.

stances are not required by all crops, but if these substances are deficient the growth of the pollen tube is slowed down and fertilization may not occur.

It is necessary that the plant have available to it micronutrients that aid the male and female organs in remaining fertile. It is also important that growth regulators such as cytokinins be available for new growth that is occurring. The grower should anticipate plant maturity and spray the plants prior to bloom with plant growth regulators to insure a healthy, viable plant at this time.

Growth Rates

That plants more or less continuously increase in size and produce new organs at least intermittently throughout their life history is one of the most self-evident of natural phenomena. The term "growth" is popularly employed to designate this complex of processes. Growth is the one plant process with which few persons are **unfamiliar,** even if they have never observed it on any larger scale than a potted plant on a window sill. However, because of the complexity of the process, the physiology of growth is still much of a mystery.

Growth almost invariably involves not only a progressive increase in the dry weight of the growing region, but a series of differentiation phenomena which becomes increasingly complex during the growth process.

Pfeffer states, "The term growth may be used to indicate all formative processes leading to a change of shape or structure." Porterfield relates that casual agents postulated as initiating, accelerating, and maintaining growth, therefore, should be studied in the light of their effects on the cell. As early as 1874 Sach postulated that the grand period of growth is due to internal factors which is exhibited by all individuals even to cell growth, these factors being influenced by external conditions.

There is evidence in plants of behavior during growth that seems to imply the effects of specific substances (hormones). Reed, many years ago, postulated with conviction of growth-promoting substances of intra cellular origin and with characteristic properties. He further stated that this "Catalyst of Growth" is abundant, but conditioned upon the rate at which organisms can obtain from the environment the materials upon which the catalyst can work. Reed concluded that it is logical to regard the problem of differentiation as a process which leads to a definite distribution of matter in space.

The growth of a plant or plant organ never proceeds steadily hour after hour or day after day, but is subject to more or less regularly recurring, often rhythmical, daily and seasonal variations in rate. A number of studies have been made of daily variations in the rate increase in the length of stems. Rose's theory of the pulsation of growth is that there is a permanent increase of growth with each pulsation resulting in an increase at night and a decrease in the day.

It is frequently desirable to give some sort of quantitative expression to the amount of growth which is accomplished by a plant or a group of plants during a given period of time. The principal indices which have been employed for this purpose are (1) increase in the length of the stem, root, or other organ of the plant; (2) increase in the area of the leaves; (3) increase in the diameter of the stem (or other organ); (4) increase in volume (especially of fruits); (5) dry weight increment, and (6) fresh weight increment.

The generalized aspects of the rate of growth of plants can often be given definite expression in graphical language. Such "growth curves" are usually plotted in terms of rate of growth or total growth increment against time. The rate of growth or growth increment may be expressed in terms of any of the commonly accepted indices of growth — elongation, enlargement phases of growth, curves plotted in terms of such units represent only the quantitative phases of growth.

If these data for the rate of elongation be plotted against time, a curve indicates that elongation is at first slow, then steadily increases until a maximum rate is attained, after which a slow but steady diminution in rate sets in, until eventually increase in the size of the cells ceases entirely. Similar curves will usually result if the rate of increase in fresh weight or rate of increase in dry weight is taken as a quantitative index of growth instead of the rate of enlargement. Every plant cell and hence every coordinated group of cells undergoes such a cyclic change in the rate of enlargement during the growth period.

That this is a universal pattern of growth behavior is

shown by the fact that it is exhibited by such diverse types of growth phenomena as the rate of elongation of segments of the root axis, the rate of elongation of segments of the stem axis, the expansion in area of leaves, the increase in weight of fruits, the growth of annual plants expressed in terms of dry weight increment, and even the growth (i.e., increase in population) of micro-organisms.

If the total increment of growth instead of the rate of growth be plotted against time, the resulting curve will assume a typical sigmoid shape. Any growth phenomenon which is represented by a curve of this type when plotted in terms of growth rates, will yield a sigmoid curve when depicted graphically in terms of growth increments. Hence, such sigmoid curves of growth are characteristic of a wide variety of growth phenomena.

Gustafson was perhaps one of the first workers to use growth curves to explain various growth events such as the location of fruits in a cluster. He found that growth varied in the tomato depending on the position of the fruit in the cluster. The first fruit to set was always the largest. His work with cukes showed that growth proceeded in very much the same manner as it does in other plants.

According to Baker and Davis the growth curves of peaches fall into two phases. The first period is the complex period due to increase in cell number and size resulting in pit and flesh increase. They state that this period is characterized by profound morphological and physiological changes. The second period or final swell being due to increase in size of cells of flesh attendant upon the rapid increase in soluble solids and water.

The growth of both normally pollinated and growth regulator induced parthenocarpic fig fruits was similar and cyclic in nature as reported by Crane. He concluded that some factor or factors, other than dominance or the reproductive tissues, controls growth of the flesh.

Dr. T.L. Senn lectures on mannose, mannitol and chelating properties of Norwegian **ASCOPHYLLUM nodosum** seaweed at National Agricultural Research Center (NARC) Islamabad, Pakistan.

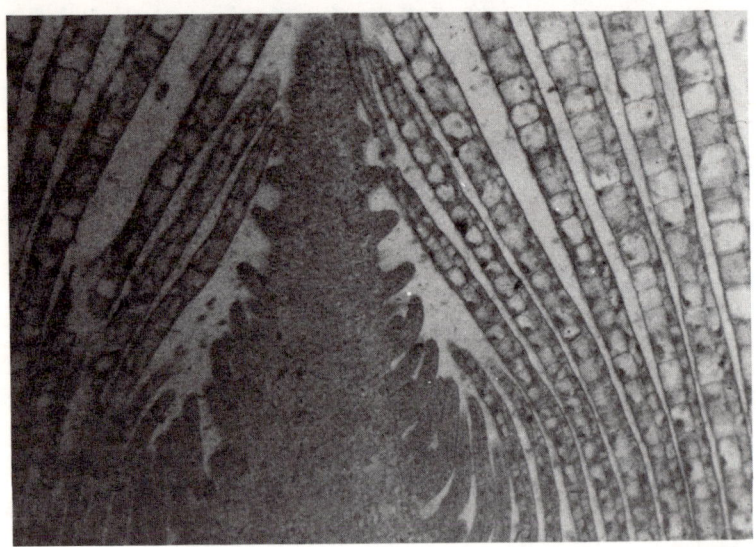

Vegetative plant bud showing cell division and elongation during rapid plant growth. Developing bud is protected by young leaves, scales and bracts.

Citrus growers use techniques involving plant growth substances in promoting bud growth by bending seedling stem at bud placement. (Courtesy of David and Polly Ellinor, Odessa, Florida.)

Left: citrus bud growth stimulated by plant growth substances. **Right:** budded tree ready to be planted in citrus grove. (Courtesy of David and Polly Ellinor, Odessa, Florida.)

CHAPTER 5
Plant Growth and Development

A good tree bringeth forth good fruit, but an evil tree bringeth forth corrupt fruit (Matthew 7:17).

Growth may be defined as an irreversible increase in size and usually an increase in dry weight (solids). These changes are measurable and quantitative. Development is reflected in changes in shape, form, and degree of differentiation. These changes are qualitative and are more noticeable than measurable.

Plant growth and development occurs in two phases, commonly referred to as the vegetative phase and the reproductive phase. During the vegetative growth phase the plant needs available nutrients such as nitrogen for growth. As a plant matures, the cells will differentiate or change and supplies of potassium and phosphorus are needed. These changes are brought about by plant regulators such as auxins. Micronutrients also play a vital role at this stage of plant development.

Agricultural practices are keeping up with changes resulting from space age research. Rapid advances in instrumentation have helped agricultural scientists solve some of the mysteries of "how plants grow." One example is in the area of plant growth regulators. Growth and development of most plant structures depend on interaction of balance of hormones.

Hormones and Micronutrients

Today the modern plant grower must speak of auxins, enzymes, hormones, gibberellins, and cytokinins. These are some of the plant growth regulators determining crop growth, development, and yield.

Equally important today are micronutrients that are part of the balance of plant nutrients controlling growth and development through their activity in enzyme systems. They are needed by growing plants in greater quantity than the soil can supply. Some of these micronutrients are iron, copper, zinc, molybdenum, boron, manganese, and cobalt. They are "micro" in the sense that only very small amounts are required by plants. Enzymes are inactive and play no specific roles until activated by a micronutrient. The micronutrients serve as starters — catalyst that activates the enzymes. Thus, each micronutrient and each enzyme plays a specific part in plant growth, development and yield. Therefore, plant growth is really dependent on available micronutrients.

Seaweed Promotes Plant Growth and Development

Seaweed has been used by plant growers for centuries, but the reason for beneficial results has only recently been attributed to the naturally occurring growth regulators and micronutrients in the seaweed.

After many years of university and private research, it has now been established that Norwegian **ASCOPHYLLUM nodosum,** and other brown seaweed, contain many naturally occurring plant growth regulators, namely cytokinins, gibberellins, and indoles. In addition, it contains essential micronutrients such as iron, copper, zinc, molybdenum, boron, manganese, and cobalt required for healthy plant growth and development. Many seaweed products also contain a chelating compound known as mannitol, which chelates micronutrients into forms that are readily available for plant use.

Featonby-Smith and Van Staden of South Africa in their publication entitled *The Effect of Seaweed Concentrate and Fertilizer on the Growth of* **"Beta vulgaris"** present the following summary:

> *The effect of foliar applications of a commercially available seaweed concentrate* **(ECKLONIA maxima)** *on the growth of swiss chard plants (Beta vulgaris L.) was investigated.* **(ECKLONIA maxima)** *at a dilution of 1:500 improved the growth of swiss chard significantly irrespective of whether it was applied on its own as a foliar spray or together with soil applications of a chemical fertilizer. The levels of cytokinin-like activity in the plants was found to be inversely related to growth. Plants treated with seaweed concentrate and liquid fertilizer had the lowest cytokinin activity in both the roots and the leaves, however, these plants grew significantly better than the controls or those which were sprayed with seaweed concentrate only. The significance of the findings is discussed.*

It was reported by Weidman and Stang that "The ability of cytokinin to induce growth and development in latent buds and to stimulate cell division offers exciting possibilities for strawberry production. The cytokinin 6-BA, alone or in combination with exogenous auxin, which is known to stimulate fruit receptacle development, could have the potential to promote development of flower buds which develop incompletely, or about prior to bloom, thereby increasing yields. Such potential to increase yields highlights the need for continued research."

Abetz and Young (1) in Australia concluded that when

Norwegian **ASCOPHYLLUM nodosum** extract was applied at the recommended rates beneficial effects in yield are obtained. In the case of the lettuce plants, seaweed extract application caused a significant decrease in the number of lettuce failing to form hearts and a significant (at the 99% confidence level) increase in the weight of marketable lettuce and in their mean heart diameter. In the case of cauliflower plants seaweed extract application led to a significant increase in curd diameter.

Many trials and experiments have been conducted in the past, the results of which have shown that seaweed extracts are responsible for increased crop yields, quality and shelf life. (See references.) Utilizing methods and instrumentation currently available to researchers, the reasons why marine-algae-seaweed extracts are beneficial to plant growth are being revealed.

The seaweed story in the 1980's is being accepted by worldwide outstanding scientists and the majority of the seaweed extracts are processed by companies with outstanding integrity. An excellent research publication by Featonby-Smith and Van Staden is an example of well planned, executed, and reported research on "the effect of seaweed concentrate on the growth of tomato plants in Nematode-infested soil." These researchers summarized their study as follows:

> *Greenhouse tests were conducted to determine the effects of a commercially available seaweed concentrate* **(ECKLONIA maxima)** *on the growth of tomato plants* **(LYCOPERSICON esculentum** *Mill.). Kelpak 66 at a solution of 1:500 improved the growth of tomato plants significantly, irrespective of whether it was applied as a foliar spray at regular intervals, or whether the soil in which the tomatoes were planted was flushed once with the diluted seaweed concentrate. Root growth*

was significantly improved whenever the seaweed concentrate was applied. Associated with this improved root growth was a reduction in root knot nematode infestation.

They discuss the significance of these findings in detail.

In 1976-1979 a series of field and greenhouse research projects were carried out in the Middle East using extracts of a brown submarine algae, Tasmanian Bull Kelp, **DURVILLEA potatorum.** It is reported that the yield of cut flower production (long-stem roses) increased by 32%. After several years of research the Israeli Department of Agriculture reported cotton yield up to 29% due to increased boll number per unit area.

Dr. Suzuki, University of Hiroshima, reported that the addition of seaweed, **DURVILLEA potatorum,** powder appeared to significantly affect the quality of tomatoes in terms of better looking fruits with less deformation. The addition of seaweed extract definitely increased the ripening and maturing rate. There are numerous additional research reports that various seaweed extracts are contributors to increase crop yields: Aitken and Senn (1965) citrus, sweet potatoes; Booth (1966) apples, strawberries; Povolny (1966) cucumbers; Blunden (1972) potato; Stevenson (1966) potato; Goh (1971) clover; Goh and Whitten (1975) clover; Blunden and Wildgoose (1977) potatoes; and others.

Seaweed products should be used as a companion material with other plant nutrients. Together they help grow better crops. They increase farm profits by increasing soil and crop health and by partially replacing the large amounts of agricultural chemicals presently used. However, good farming practices and common sense will never be replaced.

Selected References

1. Abetz, P., and C.L. Young. 1983. The effect of seaweed extract sprays derived from **ASCOPHYLLUM nodosum** on lettuce and cauliflower crops. Botanica Marina, Vol. XXVI, pp. 487-492.
2. Aitken, J.B., and Senn, T.L. 1965. Seaweed products as a fertilizer and soil conditioner. Botanica Marina 8, 144-8.
3. Aitken, J.B., Senn, T.L. and Martin, J.A. 1961. S.C. Agric. Exp. Sta. Res. Ser. No. 24.
4. Alexander, M. 1961. Introduction to soil microbiology. (John Wiley & Sons, Inc., New York.)
5. Armstrong, D.J., Burrows, W.J., Evans, P.K. and Skoog, F. 1969. Isolation of cytokinins from tRNA. Biochem. Biophys. Res. Commun., 37:451-456.
6. Blunden, G. 1972. The effects of aqueous seaweed extract as a fertilizer additive. Proc. 7th Int. Seaweed Symp., Sapporo, Japan, University of Tokyo Press, Tokyo, pp. 584-589.
7. Blunden, G. 1977. Cytokinin activity of seaweed extracts. In: D.J. Faulkner and W.H. Fenical (editors), Marine Natural Products Chemistry. Plenum, New York, pp. 337-343.
8. Blunden, G. and Wildgoose, P.B. 1977. The effects of aqueous seaweed extract and kinetin on potato yields. J. Sci. Food Agric., 28:121-125.
9. Blunden, G., Jones, E.M. and Passam, H.C. 1978. Effects of post-harvest treatment of fruit and vegetables with cytokinin-active seaweed extracts and kinetin solutions. Bot. Mar., 21:237-240.
10. Blunden, G., Wildgoose, P.B. and Nicholson, F.E. 1979. The effects of aqueous seaweed extract on sugar beet. Bot. Mar., 22:539-541.
11. Booth, C.O. 1964. Grower 62, 442.
12. Booth, E., 1966. Some properties of seaweed manures. Proc. 5th Int. Seaweed Symp., Pergamon, London, pp. 349-357.
13. Booth, E. 1969. The manufacture and properties of liquid seaweed extracts. Proc. 6th Int. Seaweed Symp., pp.

655-662. (Secretaria de la Mercante Marina, Madrid.)
14. Braid, G.H. 1978. Effects of algal extract on soil microflora. Hons Thesis, Univ. Tasmania.
15. Brain, K.R., Chalopin, M.C., Turner, T.D., Blunden, G. and Wildgoose, P.B. 1973. Cytokinin activity of commercial aqueous seaweed extract. Plant Sci. Lett., 1:241-245.
16. Briner, G.P., Richards, D., and Belcher, R.S. 1979. Seaweed products — fertilizer or plant growth regulators. N.Z.J. Agric., p. 138.
17. Brueske, C.H. and Bergeson, G.B. 1972. Investigation of growth hormones in xylem exudates and root tissue of tomato infected with root-knot nematodes. J. Exp. Bot., 23:14-22.
18. Button, E.F., and Noyes, C.F. 1964. Effect of seaweed extract upon emergence and survival of seedlings of creeping red fescue. Agron. J. 56, pp. 444-445.
19. Dekker, J. 1963. Effect of kinetin on powdery mildew. Nature (London), 197:1027-1028.
20. Driggers, B.F., and Marucci, P.E. 1964. Horticultural News, May 1964 (cited by Povolny [1969]).
21. Dropkin, V.H., Helgeson, J.P. and Upper, C.D. 1969. The hypersensitivity reaction of tomatoes resistant to Meliodogyne incognita reversal by cytokinins. J. Nematol., 1:55-61.
22. Duff, R.B., and Webley, D.M. 1959. Chem. Ind., p. 1367.
23. Fergusson, M. 1977. Liquid seaweed fertilizers. Deloraine-Westbury Ag. Topics 37, 4. (Newsl. Tas. Dept. Agric.).
24. Fogg, G.E., 1967. Nature 214, 1276.
25. Fox, J.E. 1969. The cytokinins. In "Physiology of Plant Growth and Development," ed. M.B. Wilkins. (McGraw-Hill:London.)
26. Gagnon, J.D. 1964. Nature 201, 619.
27. Goh, K.M. 1971. Kelp extract as fertilizer. I. Effect on the germination and dry matter production of white clover, N.Z.J. Sci. 14, pp. 734-748.

28. Goh, K.M., and Whitton, J.S. 1975. Kelp extract as fertilizer. II. Effect on chemical composition and element uptake of white clover. N.A.J. Sci. 18, pp. 391-403.
29. Harley, J.L. 1972. "The Biology of Mycorrhiza." (International Text Book Co., Ltd., London.)
30. Hussain, A. and Boney, A.D. 1969. Isolation of kinin-like substances from **Laminaria digitata.** Nature (London), 223:504-505.
31. Jennings, R.C. 1969. Cytokinins as endogenous growth regulators in the algae **ECKLONIA** (Phaeophyta) and **Hypnea** (Rhodophyta). Aust. J. Biol. Sci., 22:621-627.
32. Kentzer, T., Synak, R., Burkiewicz, K. and Banas, A. 1980. Cytokinin-like activity in sea water and **Fucus vesiculosus** L. Biol. Plant., 22:218-225.
33. Khaleafa, A.F., Kharboush, M.A.M., Metwalli, A., Moshen, A.F., and Serwi, A. 1975. Antibiotic (fungicidal) action from extracts of some seaweeds. Botanica Marina 18, 163-165.
34. Klambt, D., Thies, G., and Skoog, F., 1966. Proc. Nat. Acad. Sci. U.S. 56, 62.
35. Kochba, J. and Samish, R.M., 1971. Effects of kinetin and 1-naphthylacetic acid on root-knot nematodes in resistant and susceptible peach rootstocks. J. Am. Soc. Hortic. Sci., 96:458-461.
36. Miller, C.O. 1965. Evidence for the natural occurrence of zeatin and derivatives: Compounds from maize which promote cell division. Proc. Natl. Acad. Sci. U.S.A., 54:1052-1058.
37. Milton, R.F. 1963. Proc. 4th Int. Seaweed Symp., pp. 428-431.
38. Povolny, M. 1969. Einfluss des Extraktes von Seealgen auf die Lagerungsfachigkeit von Aepfein. Proc. 6th Int. Seaweed Symp., Secretaria de la Mercante Marina, Madrid, pp. 703-713.
39. Ramshaw, D. 1976. Seaweed fertilizers. Burnie District Ext. Serv. Newsl. Dec. 1976. p. 2.

40. Sawhney, R. and Webster, J.M., 1975. The role of plant growth hormones in determining the resistance of tomato plants to the root-knot nematode **Meloidogyne incognita.** Nematologica, 21:95-103.

41. Senn, T.L., Martin, J.A., Crawford, J.H. and Derting, C.W. 1961. The effect of Norwegian seaweed **(ASCOPHYLLUM nodosum** on the development and composition of certain horticulture and special crops. S.C. Agric. Exp. Res., Ser. No. 23.

42. Skelton, B.J. and Senn, T.L. 1969. Effect of seaweed sprays on quality and shelf life of peaches. Proc. 6th Int. Seaweed Symp., Secretaria de la Mercante Marina, Madrid, pp. 723-730.

43. Slade, D.A. 1967. N.Z. Fruit Growers Assoc. Summary of Expts., 1966-1967.

44. Spaull, V.W. and Braithwaite, J.M.C. 1979. A comparison of methods for extracting nematodes from soil and roots of sugar cane. Proc. S. Afr. Sugar Technol. Assoc., 53:103-107.

45. Stephenson, W.M. 1966. The effect of hydrolysed seaweed on certain plant pests and diseases. Proc. 5th Int. Seaweed Symp., pp. 405-415.

46. Terriere, L.C., and Rajadhyaksha, N. 1964. J. Econ. Entomol., 57, 95.

47. Van Staden, J. 1976. Seasonal changes in the cytokinin content of **Ginkgo biloba** leaves. Physiol. Plant., 38:105.

48. Wang, D., Hao, H. and Waywood, E. 1961. Effect of benzimidazole analogues on stem rust and chlorophyll metabolism. Can. J. Bot., 39:1029-1036.

49. Whitehead, A.G. and Hemming, J.R., 1965. A comparison of some quantitative methods for extracting small vermiform nematodes from soil. Am. Appl. Biol., 55:25-38.

50. Widdowson, J.P., Yeates, C.W. and Healy, W.B. 1973. The effect of root-knot nematodes on the utilisation of phosphorus by white clover on yellow-brown loam. N.Z.J. Agric. Res., 16:77-80.

51. Williams, D.C., Brian, K.R., Blunden, G., Wildgoose, P.B.

and Jewers, K. 1976. Plant growth regulatory substances in commercial seaweed extracts. Proc. 8th Int. Seaweed Symp. (in press).

Dr. S.M. Ismail, Director, Pakistan Agricultural Research Council; Dr. Ch. M. Anwar Khan, Director-General NARC; and Mr. Sartaj Aziz, the Minister of State for Food, Agriculture and Cooperatives, Pakistan, consider Senn's research suggestions.

CHAPTER 6
How Plants Mature

One generation passeth away, and another generation cometh; but the earth abideth forever (Ecclesiastes 1:4).

The growth and development of crop plants consists of two distinct, though overlapping, phases: the vegetative and reproductive.

The Vegetative Phase — Growth

The vegetative phase concerns essentially the development of the stems, leaves, and absorbing roots. This phase is associated with three important processes: (1) cell division, (2) cell enlargement, and (3) the initial stages of cell differentiation — change.

Cell division involves the making of new cells. These new cells require large quantities of food. Cell division takes place at the growing points of the stem and root tips. Therefore, these tissues must be provided with foods, hormones, and vitamins in order to form new cells. The plant needs magnesium, manganese, iron, boron, and copper for efficient cell division. Seaweed supplies these elements in addition to growth regulators that are active at this stage of growth.

Cell elongation concerns the enlargement of the new cells. This process requires (1) abundant supplies of water, (2) the presence of certain hormones which give the cell walls the ability to stretch, and (3) the presence of sugars. The element zinc is necessary for cell elongation.

The initial states of cell differentiation (change), or tissue formation, involve the development of primary tissues, thickening of cell walls, and development of stem and leaf systems. Growth regulators play a vital role in this cellular change.

There are two distinct phases of growth and development: preflowering (prefruiting) and flowering (fruiting). During preflowering, plants are developing their stems, leaves, and absorbing roots, and during the flowering and fruiting stage, plants are developing stems, leaves, and absorbing roots at the same time that they produce flowers and fruit. This is a very important time in the life of the plant, regardless if it is an ornamental, vegetable, or crops such as alfalfa, corn, cotton, soybean or a root crop. Utilization of carbohydrates is rapid during preflowering stages, and a balance between utilization and accumulation is necessary during the fruiting stage. Large quantities of the cytokinins, auxins, and gibberellins are needed during the preflowering stage, and a balance between these compounds and hormones are needed when the plant sets fruits and matures the fruits.

The Reproductive Stage — Maturity

The reproductive phase concerns the formation and development of flower buds, flowers, fruit, and seed or the enlargement and maturation of storage organs — fleshy stems and fleshy roots. This phase is associated with several important processes: (1) the making of new cells; (2) the maturing of the tissues; (3) the thickening of the fibers; (4) the formation of hormones necessary for the development of flower buds; (5) the development of flowers, fruit, and seed; (6) the development of storage organs; and (7) the formation of water-retaining substances.

All these evidences of maturity require a supply of stored food. In most cases these foods are the starches and the sugars.

The vegetative-reproductive balance and the type of growth and use of the food supply depends on many factors, such as water supply, temperature, light supply, and available nutrients. In general, all crops require a dominance of the vegetative phase during the germination and seedling stages. This is an initial stress period for the plants and it is essential that growth promoting substances be available at this time. Growth regulators play a vital role in the change from a vegetative plant into a reproductive or maturing one. Accordingly, large quantities of the cytokinins, auxins, and gibberellins are needed during the spring and early summer, and relatively large quantities of the florigens, or growth inhibitors, are needed during the late summer and fall.

The micronutrients are necessary for cell division and cell enlargement and marine algae-seaweed in foliar sprays supplies the growth regulators needed for changes associated with maturity. Time of application is very important in the application of all foliar sprays.

Essential Elements — Limiting Factor

As growers well know, certain factors of the environment and certain cultural factors greatly affect the two phases of plant growth and development. The most important factors are (1) the water supply, (2) the temperature, (3) the light supply, and (4) the nutrient supply. These factors usually limit the growth and development of plants and they are called the limiting factors in plant production.

It is now understood that the growth and final yield of any crop are regulated largely by a limiting factor. Sometimes the carbohydrate supply, sometimes the nutrient supply or a particular element, sometimes the water supply, or sometimes the temperature are limiting factors alone or in combination.

Quite often the solving of the grower's problems consists in discovering and correcting the limiting factor.

Since the limiting factor most likely determines the rate of growth and yield, an appreciation of the role that marine algae-seaweed can play in correcting the limiting factor is necessary.

The following chapters will illustrate how seaweed can supply the plant with micronutrients and aid the plant in overcoming stress.

The green plant is a biochemical factory. Certain raw materials are used, either directly or indirectly, in the making of all-important foods, fibers, enzymes, hormones, and vitamins. These raw materials meet at least two requirements: (1) they contain one or more essential elements for growth and development, and (2) they exist in a form which plants can absorb and use. For example: nitrogen is a part of the chlorophyll molecule and essential for growth. Potassium is usually associated with maturity and reproductive processes. As previously pointed out, micronutrients are needed throughout the life of the plant and enter into the formation of plant growth substances.

Mature-Maturation-Maturity-Ripe

Mature is to become fully developed or as a grower usually says — "ripe." Maturation to the researcher is having completed natural growth and development, or in other words, the final stages of differentiation of cells, tissues, or organs. Maturity then is the quality or state of full development or termination of a period of life.

Good Farming Practices Needed

Growth and maturity of plants depend on good cultural practices, correct timing of foliar sprays, use of soil testing, and following instructions regarding the use of micronutrients and growth promoters.

In summary: An old proverb to keep in mind. As long as you are green you grow, but once you think you are ripe (mature) you begin to rot.

On a recent trip to the British Isles, Dr. T.L. Senn observes plant growth and development in Scotland.

CHAPTER 7
Why Plants Need Micronutrients

Everything in nature contains all the powers of nature, everything is made of one hidden stuff (Emerson, Compensation).

Biochemical Reactions in Crop Plants

The green plant is a biochemical factory. Certain raw materials are used, either directly or indirectly, in making of the all-important foods, fibers, enzymes, hormones, and vitamins.

Photosynthesis is essentially an energy-fixing reaction. As is well known, this energy comes from the sun in the kinetic form and is changed to the potential form found in foods. Crop plants, the only connecting link between ourselves and the life-giving sun, assume great practical significance.

In contrast to photosynthesis, respiration is essentially an energy-releasing reaction. The potential chemical energy of foods is transformed into various kinds of kinetic energy. In this way, the light-energy reserve built up by crop plants becomes available, not only for the crops themselves, but also for all mankind. Photosynthesis fixes, or stores, the free energy of the sun, whereas respiration releases it.

Essential Elements and Essential Raw Materials

The essential elements necessary for plant growth and development, namely nitrogen, phosphorus, potassium, sulfur, calcium, and magnesium are well known. Extensive research has been recorded pointing out their importance in the growth of plants. They may also be referred to as essential raw materials needed in the manufacturing of foods such as carbohydrates and oils by plants.

These raw materials meet at least two requirements: (1) they contain one or more essential elements for growth and development, and (2) they exist in a form which plants can absorb and use. For example, nitrogen is part of the molecule of all proteins and part of the molecule of both chlorophyll a and chlorophyll b (the green coloring matter in plants). Nitrogen is, therefore, an essential element. Although nitrogen exists in many types of compounds, crop plants absorb and use nitrogen from the soil mostly in two relatively simple forms: the nitrate ion and the ammonium ion. Because these ions are absorbed and used by plants in making nitrogenous organic compounds, they are called essential raw materials. Thus, essential raw materials are chemical compounds or parts of chemical compounds which contain one or more essential elements for plant growth and are absorbed and utilized by plants.

Micronutrients

The essential micronutrients necessary for plant growth and development are manganese, iron, boron, zinc, copper, molybdenum, and cobalt. Since any one of these elements may become a limiting factor in growth and development, the grower should learn what the role of these elements is in plant life, how to recognize symptoms of micronutrient element deficiency, and when and how best to replenish the supply. Soil and plant tissue tests are great aids in determining crop plant needs.

There are micronutrients in the soil that were present in the original rocks from which the soil was formed. Soils in hot, moist areas contain smaller amounts of micronutrients than soils in cold, dry areas. Even though micronutrients are present in most soils, some of them are bound up in soluble organic acids and are not available for plant growth. The availability of micronutrients for plant use is determined by their solubility, which is dependent upon the alkalinity or acidity of the soil solution.

Micronutrient deficiencies often occur at a very critical stage in the life of a plant. For example, zinc plays a vital role in the fertility of female flowers of nut trees. Soil applications of zinc sulfate are used, but if the deficiency occurs during preflowering the grower often resorts to foliar applications for quick response. Plants can absorb micronutrients through their leaves and stems and most micronutrient deficiencies can be temporarily corrected by foliar applications. It is wise to explore the cause of the deficiency and try to correct the problem by soil applications.

Role of Micronutrients

Manganese is part of an enzyme system for condensing soluble amino compounds into proteins. Iron has two roles: as part of an enzyme system for making of chlorophyll and as part of the molecular structure of the heme pigment needed for respiration.

Boron is necessary for cell division, the germination of pollen, the movement of sugars through membranes, the development of conductive vessels, and the transport of certain hormones. Boron prevents internal tissue breakdown which is evidenced by cracked stems.

Zinc is necessary for cell elongation, for the formation of protein, and for the oxidation phase of respiration.

Copper is an essential constituent of certain oxidizing enzymes. It is necessary for the formation of the green coloring matter of plants — chlorophyll.

Molybdenum is part of the molecular structure of an enzyme which turns on the nitrate reduction system. Crop plants are unable to utilize nitrogen without molybdenum. It is very important to the nodule bacteria of legumes.

Micronutrients have many functions in crop plant growth and development. The amount and availability of micronutrients will vary with soil types and the demand by different crops. Even though the amounts required by plants are small, the micronutrients are just as essential as the nitrogen, phosphorus, and potassium, and in some areas even more so.

Specialty products are especially developed to supply the plant with micronutrients that in combination with marine alga such as **ASCOPHYLLUM nodosum, DURVILLEA potatorum** and **ECKLONIA maxima** make available to plants usable elements. Micronutrients are necessary for plants in times of plant stress, such as flowering, maturing, and during periods of drought.

Seaweed is a source of iron and chelating compounds such as mannitol. There has been much discussion on the amount of micronutrients in seaweed and seaweed extracts. The amount of micronutrients required by plants varies with the particular crop and the soil in which it is grown. Seaweed contains chelating compounds which will make some micronutrients already in the soil (but which are not available to plants because of solubility) available to the plants in a chelated form.

These claims are supported by the research of Lynn (1972) and Aitken (1974) who added several concentrations of seaweed extract to mineral deficient nutrient soils. Franki (1960) found leaves of tomatoes treated with seaweed extract contained more manganese than was present in the seaweed itself. He concluded that the seaweed had released "unavailable" manganese from the soil.

Abetz and Young (1983) reported that lettuce plants harvested were not significantly affected by NPK side dressing, but seaweed extract treatment increased the yield significantly (at the 95% confidence level). Similarly the standard yield per plot was not significantly affected by NPK side dressing but was significantly increased by seaweed extract treatment.

The mean heart diameter of all lettuces was significantly increased by all three seaweed extract treatments but not by the NPK side dressing.

Abetz and Young concluded that seaweed extract assists the lettuce plants to more efficiently utilize the major elements present in the soil.

Seaweed has been used for centuries in the United Kingdom. Senn visits a rose grower in England that has been using seaweed products for many years.

Selected References

1. Aitken, J.B. 1964. The effect of seaweed extract and humic acids on the O_2 uptake of **Citrus sinensis** seedlings grown in nutrient deficient culture. M. Sc. Thesis, Clemson University.
2. Abetz, P. and C.L. Young. 1983. The effect of seaweed extract sprays derived from **ASCOPHYLLUM nodosum** on lettuce and cauliflower crops. Botanica Marina, Vol. XXVI, pp. 487-492.
3. Frank, R.I.B. Studies of manurial values of seaweeds. Plant Soil, 1960. 12,311.
4. Lynn, L.B. 1972. The chelating properties of seaweed extract **ASCOPHYLLUM nodosum** vs. **MACROCYSTIS pyrifera** on the mineral nutrition of sweet peppers **CAPSICUM annum.** M. Sc. Thesis, Clemson University.

New Zealanders love their homes and gardens and are users of seaweed. The author has visited with growers and homeowners on several recent trips to New Zealand.

CHAPTER 8
Plant Growth Regulators

Things do not happen in this world — they are brought about (Will Hays).

Plant hormones are classified upon the basis on the kinds of response they influence. The bending of plants to light sources (phototropism) and the influence of roots going down and tops coming to the surface (geotropism) represent two of the many hormone influenced responses of plants. It is highly evident that all developmental processes are influenced to some degree by one or more kinds of hormones. Research has shown that hormones, auxins, and growth regulators are involved in the growth and development of roots, stems, and leaves. Stress periods occurring prior to and during formation of buds, root hairs, tubers, flowers, and in fruit development and maturity, senescence, dormancy of buds and seeds and germinating of seeds are periods of great growth regulator activity. Some seaweed contain many of the known natural growth regulators. The growth and development of plant parts evidently is the result of interaction of many of these growth regulators.

Enzymes

In crop plants, thousands of chemical reactions are taking place at the same time. Enzymes consist of proteins, with or without some other compound, and are necessary for these reactions to take place. Enzymes are effective in minute concentrations and are specific in action. For example, the carbohydrases work on carbohydrates only, the lipases react with the oils and fats only, and the proteases react with proteins only. This specificity may be compared to the indentations of a specific key fit into a lock for which the key is made. In this way, the numerous chemical reactions within crop plants are more or less controlled.

Hormones

Hormones are similar to enzymes in that they are effective in very low concentration, though they differ from the enzymes in that they are less numerous than enzymes and are made in one part of the crop plant and translocated to another. Hormones promote or stimulate crop growth and at certain times suppress vegetative growth.

Hormones which promote or stimulate growth and development are the cytokinins, the auxins, and the gibberellins. In general, the cytokinins are made in the region of cell division of the root and are translocated to the region of cell elongation of the stem, where they seem to be necessary for the making of new cells. Seaweed extracts contain many of the naturally occurring cytokinins and when applied as a foliar spray is taken into the leaf and translocated to the active regions.

The auxins are made in the regions of cell division of both root and stem and are translocated to the region of cell elongation, where they give the cell walls the ability to stretch. The gibberellins are made in active leaves and are translocated in the conducting vessels to the regions of cell elongation, where with the auxins, they facilitate cell elongation.

Practical Applications

The cytokinins made in the roots, and the auxins and gibberellins made in the tops interact, or work together, in the development of crop plants. Research has shown that these three types of hormones are needed for the profitable production of crops.

Some seaweed extract products are especially formulated to supply plants with additional hormones at critical stress periods. When cytokinins are applied to leaves, these treated leaves remain green longer due to a slower rate of degradative processes, and syntheses are stimulated. In other words, the treated leaves are rejuvenated and manufacture foods for a longer period of time.

The activity of the female flower is stimulated by various micronutrients and growth regulators. This insures a better

Business and pleasure are combined on trips to Australian farms and greenhouses.

set of fruits and seed pods. Many investigations have shown that the application of growth regulators promotes the development of roots. These materials not only speed up the healing of cut surfaces, but they also induce the development of a large number of roots.

Cytokinins, when applied to apple trees, promote the growth of the spurs and the development of lateral branches. Growth regulators also reduce preharvest dropping of fruits.

Bioassay Systems to Test for Plant Growth Hormones in Extracts of ASCOPHYLLUM nodosum, Algae

Alta R. Kingman and T.L. Senn
Clemson University
Clemson, South Carolina 29634

Abstract

When certain species of marine algae are dried, ground, and added to media in which plants are growing, or are sprayed onto the foliage as aqueous suspensions, the effects on plant growth often exceed that which can be explained by the known chemical composition of the seaweed. The objective of this research was to measure auxin, cytokinin, and gibberellin-like responses in **ASCOPHYLLUM nodosum** extracts using bioassay techniques.

Three standard bioassay systems were used to test for plant growth hormones in extracts of **ASCOPHYLLUM nodosum**. The **Avena** straight growth test showed that seaweed concentrations of 100 μg/ml elicited auxin-like responses from coleoptile segments. The tobacco callus tissue test for cytokinins revealed significantly greater proliferation of callus than the control or the kinetin standard when treated with 10 μg/ml, 50 μg/ml, or 100 μg/ml of **ASCOPHYLLUM nodosum** extract. The measurement of gibberellin-induced a-amylase release from barley endosperm was used to measure gibberellin-like substances in **ASCOPHYLLUM nodosum**. Significant increases in amylolytic activity over the control and the gibberellic acid (GA$_3$) standard with seaweed concentrations of 10 μg/ml, 100 μg/ml, 500 μg/ml or 1000 μg/ml indicated the presence of gibberellin-like substances in a dehydrated powder of **ASCOPHYLLUM nodosum**.

Although the seaweed extracts tested herein demonstrated positive responses in the GA$_3$ sensitive system, it is not yet known whether high concentrations of this gibberellin, or others, or combinations exist in the seaweed material assayed. It would be worthwhile to employ gas chromatographic and/or other bioassay screening and identification techniques.

A positive growth response, such as that exhibited by extracts of **ASCOPHYLLUM nodosum**, from tobacco callus tissue reflects the presence of cytokinin-like substances. Replicate cultures were very much alike within a given treatment both in size and appearance, although some variability was observed in total yields. This, in part, can be attributed to some lack of uniformity of the planting material, variability in growing conditions over the period of time during which the research proceeded, and variability in tissues used from different plants as well as difficulty in obtaining tissue from precisely the same place and of the same size on stems.

It is significant that the seaweed extracts reported herein demonstrated each time a well-established response to plant growth regulators; e.g., decreasing or inhibitory responses with increasing concentrations. Additionally, results reported herein show that the extracts of **ASCOPHYLLUM nodosum** do contain auxin, gibberellin and cytokinin-like substances.

Response of **ASCOPHYLLUM nodosum** Extracts in the **Avena** Straight Growth Test

Worldwide research on the agricultural uses of seaweed has suggested growth regulatory components of some type in extracts of **ASCOPHYLLUM nodosum**, a brown alga. When the material is applied to plants as a foliar spray or incorporated into the growing media as a meal, it regularly results in growth effects that cannot be explained on the basis of its known nutrient composition.

The research reported herein was conducted as an initial effort to determine the presence or absence of auxin or auxin-like substances in a wettable powder of **ASCOPHYLLUM nodosum** using the **Avena** straight growth bioassay.

When concentrations were high (above 100 μg/ml), the seaweed extracts exhibited the classical response of growth regulators; decreased growth with increased concentrations. The material used in this research has been assayed after its incorporation into commercial preparations. Williams, et al., were not able to demonstrate auxin activity in several commercial extracts of **ASCOPHYLLUM nodosum** using wheat coleoptile bioassays.

The **Avena** bioassay used in this research is specific for auxins, although not as sensitive as the **Avena** curvature bioassay. Concentrations of seaweed extracts (100 μg/ml) elicited auxin-like responses; therefore, the presence of indole compounds is indicated. Care must be taken not to extrapolate from the tissue system used here in an attempt to relate these concentrations to the intact plant, although effects of plant growth and development appear to be similar to responses elicited by low concentrations of indole compounds as well as other known growth regulating compounds.

The research findings herein indicate that **ASCOPHYLLUM nodosum** extracts at levels of 100 μg/ml elicit auxin-like responses from **Avena** coleoptile segments. Gas chromatographic techniques using the concentrations indicated by these research findings would be more definitive. Indole derivatives and quantities present in the seaweed must be established in order to ascertain whether purported responses of growing plants to **ASCOPHYLLUM nodosum** applications are due to auxin-like compounds in the seaweed.

The Gibberellin and Cytokinin Activities of **ASCOPHYLLUM nodosum**

The objective of the research reported herein was to measure cytokinin- and gibberellin-like responses in **ASCOPHYLLUM nodosum** extracts using bioassay techniques.

Results

The pattern of response in a test for gibberellin-like substances in **ASCOPHYLLUM nodosum** extracts was similar to that of a-amylase release from barley endosperm by gibberellic acid (GA_3). The amylolytic activity of the seaweed was also proportional to the logarithm of the concentration of GA_3 used for the standard of comparison.

Sufficient clean white tobacco pith callus tissue was obtained for all subcultures. All seaweed treatments resulted in significantly better growth of callus tissue than the control or the kinetin standard. Seaweed extract at 10 μg/ml resulted in the greatest proliferation of callus.

It is significant that these seaweed extracts reported herein demonstrated each time a well-established response to plant growth regulators; e.g., decreasing of inhibitory responses with increasing concentrations. Additionally, results reported herein show that the extracts of **ASCOPHYLLUM nodosum** do contain gibberellin- and cytokinin-like substances.

Additional Research Reports

Kingman and Moore at Clemson University devised a quick and accurate method of quantitating a broad range of plant growth regulating substances. They stated that following extraction and filtration, gas liquid chromatography (GLC) recoveries were 65% for ABA; 80% for adenine; and 60% for IAA, indicating suitability of their methods for GLC determinations of these plant growth substances in aqueous solutions of the brown alga, **ASCOPHYLLUM nodosum**. Their research did not include investigations into gibberellin presence in seaweed extracts; but, this would seem to fall into the logical process.

Tay, MacLeod, and Palni in Australia (1985) report that **t**-zeatin, **t**-zeatin riboside, their dihydro derivatives, isopentenyladenine and isopentyaldenosine were identified and quantified in a commercial extract of Tasmanian Giant Bull Kelp, **DURVILLEA potatorum.**

Gas liquid chromatography (GLC) is now the accepted method of analysis of plant extracts for cytokinins. Many bioassay techniques have been reported which are less specific, but the equivalent kinetin level (EKL) may be established instead of identifying and quantifying a particular cytokinin. GLC is the preferred method in that it can be used to determine which particular substances are present and in what concentrations.

Young, reporting from Australia, stated that the major plant growth regulators in **ASCOPHYLLUM nodosum** are almost certainly cytokinins. He supports this statement with bioassays and chemical analyses. He further comments that there is some doubt as to the identity of all the cytokinins present in seaweed extracts, but gas chromatographic analysis suggests that adenine, (6-aminopurine), 6-(Y-Y-dimethylallylamino) purine, kinetin, (6-furfurylaminopurine) and zeatin and possibly 6-benzylamino purine are present. Adenine and zeatin are the major cytokinins present. Adenine exhibits low biological activity whereas zeatin is the most active cytokinin known.

The seaweed extract prepared from **ASCOPHYLLUM nodosum** has an equivalent kinetin level (EKL) of not less than 50 $\mu g/g$ of dry extract.

Naturally occurring cytokinins, with perhaps the exception of kinetin, are stable over long periods of time provided they are not subjected to high temperatures or to high light intensity.

Menary, of the University of Tasmania, did soybean callus bioassays on 1976 and 1977 **ASCOPHYLLUM nodosum** commercial extracts in 1977 and reported the kinetin equivalent of the two samples to be the same.

Future Outlook

Plant growth regulators have been in use for well over fifty years. Many growers use plant growth regulators without knowing what they are and why they are being used. This action is rapidly changing and the growers of the future will ask questions, and rightly so.

A review of the scientific literature will reveal that plant growth regulators are being widely researched the world over. Plant growth regulators are finally receiving the attention on the same level as pesticides, herbicides, and fertilizers.

Some of the most promising potential for the future application of new research in agriculture is the use of plant growth regulators. Several of these materials are already available to the progressive grower. Much remains to be researched and much is to be learned about the mode of action of the various natural plant growth regulators. As of now, most of the available plant growth regulators can be classified as auxins or auxin-related materials, gibberellins, cytokinins, or growth retardants.

The field of natural plant growth regulators is rapidly changing and there is much promise for biological agriculture.

Effect of increasing concentrations of plant growth regulators in **ASCOPHYLLUM nodosum** on the fruiting of pepper **(CAPSICUM).**

CHAPTER 9
Seaweed and Plant Stress

He that endureth is not overcome (Proverb).

Stress

Call it strain. Call it pressure. Call it tension. By any name, stress is the reaction of our bodies to any threat or change. There is no such thing as a stress-free environment for man nor plant.

A review of the highlights of stress as it relates to our own lives may help us to have a stronger appreciation of plant stress.

Most of us sometime in our growth and development have encountered stress. We are aware that regardless of the basic personality pattern of the hypertensive person, blood pressure elevation usually is aggravated by emotional stress.

From allergies and asthma to cancer and heart attack, research suggests that high levels of stress help promote illness by changing the immune system and placing an added load on the heart and blood vessels.

Later, we will review the subject of seaweed, hormones, and cytokinin effects on plant growth. First, let's investigate the effects of stress hormones on our own bodies.

Stress hormones are released into the body when the brain receives the signal that danger is near. The hypothalamus releases the hormone known as corticotropin (CRH). The CRH is released near the pituitary gland. This sets up a chain reaction in the body. The pituitary in turn releases a chemical called adrenocouticotropic hormone (ACTH). It travels through the blood to the adrenal glands. The adrenal glands then produce cortisol which activates glands throughout the body to prepare for stress. It is reported that this whole process takes just 30 to 45 seconds. This example is cited to impress on the reader how fast biochemical reactions may be.

Even though this manuscript is written about plants, it is important that the reader have an appreciation of human stress. Some very important understanding of the effects of stresses on plants have come from research of animal cells.

Plant Stress

It is very difficult to define plant stress in simple everyday terms. Growers have always been very concerned about the influences of changes in the environment on plant growth and development. Levitt (1980) states that "Biologists have adopted the term **stress** for any environmental factor potentially unfavorable to living organisms and **stress resistance** for the ability of the plant to survive the unfavorable factor and even to grow in its presence."

The plant may be considered as an integrated mechanism capable of an irrevocable increase in size and complexity. Starting with the germination of a seed, the developmental history of a plant may be traced through juvenility, maturity, flowering, and fruiting. At fruiting the essential cycle of the plant growth is completed.

As with man at each major change in growth and development, such as puberty, visual evidence is encountered. Plants produce buds or flowers as evidence of the state or quality of being first capable of begetting or bearing offspring. Prior to this, tremendous changes have occurred in the plant cells and tissues. Changes have been motivated and/or accelerated by growth hormones or growth regulators. These regulators were dependent on elements such as those found in seaweed and are required at specific times in the life cycle of the plant.

The first stress period in a plant's life is just after germination. Germination includes all the sequential steps from the time the seed imbibes water until the seedling is self-sustaining. In the simple concept, germination involves the enzymatic conversion of complex reserve foods to simple soluble substances that are readily translocated to the young plant. Thus, starter solutions and seaweed adjacent to the young roots in the soil are very beneficial to the plant during this period of stress.

Seed Physiology

It has been previously reported that seaweed products might hasten the germination of seeds, an effect which may be attributed to the auxin or growth hormone present in seaweed.

Based on these reports, research was initiated at Clemson to investigate the effect of various extracts on the subsequent respiratory activity of seed and their ultimate germination potentials. Beet seed were used to determine the effect of seaweed extract on germination. At the end of one week the germination of the treated seed was 84% as compared to 0% for the checks. Another experiment was conducted to study the effects of soaking seed in seaweed extracts for 30 minutes prior to germination. Prior soaking of seeds for 30 minutes increased the germination of beet seed by 25% over the control.

Treatment of several species of seed greatly accelerated the respiratory activity of the seed. The germination percentage of the treated seed was also increased. Five milliliters of seaweed extract was applied to the seed in a petri dish in varying concentrations. The concentrations varied from the pure extract through 1 to 5, 1 to 10, 1 to 25, 1 to 50, 1 to 100, 1 to 200, 1 to 300, 1 to 400, and 1 part seaweed extract to 500 parts of water.

In all cases tested, application of the extract, even in low concentrations, increased the respiratory activity of the seed. The higher the concentration of extract, the higher the rate of respiratory activity. In the higher concentrations, where the respiratory activity was high, the germination percentage was low. However, when the concentration was such that the respiratory activity was only moderately increased, the germination percentage was increased. The optimum concentration varied considerably with different species of seed. Seaweed extract in these tests contains the potential to increase germination of certain seed.

Environmental Stress

Man has long been interested in understanding the environmental limits for the existence of life on earth. We have also been interested in understanding the mechanisms within plants that establish these limits and how these mechanisms might break down under the stresses caused by environmental extremes.

There are several situations in which the stresses occurring at the extremes control crop yield.

Drought — Water is clearly one of the most important environmental factors from the standpoint of the plant's function. In very recent years, because of important advances in the physics and chemistry of water, a better understanding of the stress factor on plant growth has been encountered. Applications of seaweed extract during early drought periods aid plants to live through this stress period.

One of the first results of drought — wilting of plants — is a decrease in growth promoters and an increase in growth retardants. There is considerable literature available on growth retardants, ABA, PA, and DPA. Growth promoters in relation to drought stress are not as well researched.

Itai and Vaadia (1968) reported that a decrease in cytokinin translocation from the roots of plants with increased osmotic stress. As the osmotic stress was returned to normal, the cytokinin activity was restored to normal. They concluded that the stress induced decline in protein synthesis in the leaves is due to a deficiency of cytokinins.

Some plants that withstand drought are able to initiate stomal closure much more rapidly in response to water stress. In stressed plants the cytokinin decreases in quantity (Itai and Vaadia, 1965) and it stimulates transpiration. On the other hand, abscisic acid reduces transpiration and produces stomatal closure (Little and Eidt, 1968).

Excess water also results in wilting of plant leaves. Burrows and Caar, 1969, found that flooding decreased the synthesis of cytokinins in root tips of sunflower and cytokinin export from root to leaves was reduced. Foliar sprays of benzyladenine to tomato leaves overcame the dwarfing effect of excess water, Reid and Railton, 1974. Wright and Hiron, 1972, stated that flooding-induced chlorosis was due to a reduction in transfer of cytokinins from the flooded roots to the shoots.

High temperatures — High temperatures typically accompany drought conditions and are an important environmental stress factor in themselves. It is believed that high temperatures and drought conditions have a direct effect on the breakdown of plant enzyme systems. Research has shown that applications of seaweed extracts will aid the plant during this type of stress. The reasons are not clearly understood but may be attributed to the micronutrients being readily available and to the growth regulators (cytokinins) supplied by the seaweed.

Shariff and Dale (1980) reported that under conditions of mineral nutrient stress assimilate supply to the shoots is reduced restricting the nutrients available for growth as well as the cytokinin supply from the roots. The practical application was that tiller bud growth in barley could be increased with the application of cytokinins to the roots of plants under nutrient stress and that the increase was greater if plants were supplied with both cytokinins and mineral nutrients.

Scientists at the Newton Research Laboratories in Australia reported that seaweed extracts increased the drought resistance of plants under stress. They found that by reducing the transpiration rate the plant loses less moisture under drought conditions resulting in less damage to water stress. Seaweed extracts evidently aid in changes in the metabolic pathways thereby permitting the roots access to extremely low moisture levels which during a drought are normally not available to plants.

Nelson and Van Staden (1984) found that a commercially available seaweed extract **(ECKLONIA maxima)** when applied to nutrient-stressed plants of cucumber as a root dip at transplant or as a weekly foliar spray resulted in overall plant dry mass increase. The plants receiving the highest levels of seaweed treatments showed greatly increased root growth. They concluded that the results on cucumbers suggest that seaweed extract may have increased uptake of "unavailable" nutrients by cucumber roots, or has improved efficiency or utilization of "available" nutrients. Seaweed extract treatment may be expected to increase plant growth, even when the plant is under nutrient stress.

It has frequently been reported that plants treated with various plant growth regulators were less susceptible to environmental stress conditions such as water (drought), temperature (frost), and diseases.

Flowering

Flowering is a term representing a wide spectrum of physiological change and form and structure of plant events. The first event is the most critical, it is the change from stem to flower. At this time biochemical changes take place that alter the structure of the plant.

The second major change or stress period occurs when the plant is changing from a vegetative to a reproductive plant. Prior to this change or prior to blooming is an ideal time to apply foliar applications of seaweed extract. The seaweed extract supplies the plant with readily available micronutrients and growth regulators that are so vital for healthy male and female flower parts.

The importance of flowering in the growth and development of all plants is often overlooked. The floricultural industry is largely dependent on the control of flowering in plants. In some crops the inducement of flowering is vital in meeting market demands such as lilies at Easter, poinsettias at Christmas and the demand for other flowers on special occasions. Likewise, the prevention of flowers is very important in crop plants when flowering is undesirable.

The altering or changes in plant growth habits always results in a period of stress. Flowering in plants may be said to be the end result of all growth and developmental activities.

Many research publications have shown that plant growth regulators influence plant growth, development of buds and fruiting in many crops.

The stimulation of bud development in many plant species has been attributed to the influence of cytokinins. Gibberellins have been known to induce flowering in many plants.

Sachs (1967) reported on the release of auxiliary buds from correlative inhibition in intact plants by applications of a cytokinin.

Weidman and Stang (1983) stated that the ability of cytokinin to induce growth and development in latent buds and to stimulate cell division offers exciting possibilities in strawberry production.

Plant Response to Stress

Plant response to stress is highly complex. It could involve specific enzymes. There are also many other stress factors that have not been covered. The rate of transpiration (water loss) may have a great bearing on plant stress.

Much research remains to be conducted in the study of how plants recover from periods of stress.

Stress References

1. Itai, C. and Vaadia, Y. 1965. Kinetin-like activity in root exudate of water stressed sunflower. Physiol. Plant. 18, 941-944.
2. Itai, C. and Vaadia, Y. 1968. The role of root cytokinins during water and salinity stress. 1st T.J. Bot. 17, 187-195.
3. Levitt, J. 1980. Responses of plants to environmental stresses. Academic Press, NY.
4. Little, C.H.A. and Eidt, D.C. 1968. Effect of abscisic acid on bud break and transpiration in woody species. Nature (London) 220, 498-499.
5. Reid, D.M. and Railton, I.D. 1974. The influence of benzyladenine on the growth and gibberellin content of shoots of waterlogged tomato plants. Plant Sci. Lett. 2, 151-156.

6. Wright, S.T.C. and Hiron, R.W. 1972. The accumulation of abscisic acid in plants during wilting and under stress conditions. Plant Growth Substances, 1970. Springer-Verlag, NY.

Effect of varying concentrations of Norwegian **ASCOPHYLLUM nodosum** seaweed extract on top and root growth of coleus.

CHAPTER 10
They Condemn What They Do Not Understand

He that would rightly understand a man, must read his whole story (Proverb).

We have all heard someone say, "It won't work; I don't understand it." To understand is what is hard. Once one understands, action is easy.

Few of us truly understand television, space ships, or even how plants grow, but most of us watch television, read about space travel, and growers still plant seed.

There is no unbelief; whoever plants a seed beneath the sod and waits to see it push away the clod — he trusts in God (L.Y. Case — Unbelief).

Whether I am on the winning or losing side is not the point with me; it is being on the side where my sympathies lie that matters, and I am ready to see it through to the end (Seeger — Letter From the Trenches).

My sympathies lie with the grower of plants. My enthusiasm is for new ways to help the grower find a better way. As we find new ways to make progress in the basic understanding of the biochemistry of plant growth and development there will be a better way.

Biological farming, after all, refers to life, living, growth, and development. Attitude, belief, and understanding may be the difference between success or failure of a crop. That alone is reason enough to read the whole story.

Basics

However, let's get down to basics, or as some would say, the "nitty gritty." We know that all plants must carry on the fundamental processes of photosynthesis and respiration in order to grow and develop. These processes require micronutrients and growth regulators for their successful initiation and completion.

Seaweed supplies these much needed micronutrients at a time when the plant requires them. We also know that for growth to take place the plant is dependent on growth regulating substances. Seaweed, an important ingredient of starter solutions, supplies many of the naturally occurring plant cytokinins. All of these growth regulators are so vital for healthy plant growth.

The growth process involves cell division and cell enlargement. Micronutrients are necessary for both of these processes. Cells also stretch or must have flexibility and micronutrients give cells this activity.

Development

Cells mature bringing about great changes in the plant. Evidence of these changes is formation of flower buds which is a sign of reproductivity. Growth regulators are very important prior to and during this vital stage in plant development. This is an excellent time to foliar apply seaweed extracts. The micronutrients are in a chelated form and readily available to the plant. The cytokinins and other growth regulators are absorbed into the leaves and at the place of need.

The flowers represent the sexual mechanisms of the plant. It is very important that the female and male parts be receptive at the same time. The male must be potent and pollinate and fertilize the female. Most micronutrients are used by the plant during this stage of plant development.

Why use seaweed? Seaweed extract aids the plant in development, which puts more fruit on the tree, more pods on the vine, more grains on the cob, and, of course, more money in the pocketbook.

Adverse Conditions

Many adverse conditions occur during the lifespan of crop plants. Perhaps drought is the most common and perhaps the most harmful. Water determines cell size and also keeps the leaves in a manufacturing condition. Most plant parts are 80%-90% water.

High temperature is usually associated with dry weather and the plant wilts. Also, at high temperatures plants utilize more foods and, therefore, have less for growth and storage.

Research has shown that micronutrients and growth regulators help a plant withstand the problems caused by stress. Therefore, a prime reason for using seaweed is to help overcome stress problems.

Micronutrients

The Good Book says that man cannot live by bread alone, likewise plants cannot live by N-P-K alone, both need balanced nutrition. Biological farming helps the grower to have balanced nutrition available for his plants. Seaweed supplies the plant with growth regulators and micronutrients not available otherwise.

HOW COULD IT HAPPEN?
Seeds

Many researchers have reported speeding up seed germination by applications of seaweed extract. Senn records variations in the respiratory rate of the plant under tests which are related to the effect of micronutrients in the seaweed on enzymic activity in the plant. Enhanced seed germination was recorded on all of 12 species tested. Similar reports have been made by Simpson, who stated that the percentage emergence was increased in each of the four soils tested.

Biddington and Thomas (1976) investigated the influence on germination of lettuce and celery seed of various cytokinins. All the cytokinins were active in promoting germination of lettuce seeds. Reynolds and Thompson (1973) found that kinetin at concentrations of 0.1 to 10 ppm strongly promoted the germination of lettuce seeds effectively removing the inhibitory effect of high temperatures on germination. Sharples (1973) made a similar study at 30° C and 35° C and found that kinetin stimulated lettuce seed germination. Esashi and coworkers (1975) found that 6-benzylaminopurine stimulated germination of cocklebur seeds; Kuribayashi and Ohashi (1975) found that kinetin could shorten the germination time of Panax ginseng seeds by half if seeds were treated with 75 ppm kinetin solution for four days.

Cook (1982) reported that using a Norwegian **ASCOPHYLLUM nodosum** seaweed extract as a pregerminating medium for onion seeds treatment at 10° C demonstrated the greatest variation from the control and a positive correlation was seen between concentration and germination rate. The seaweed extract was temperature specific and gave an increase in germination as well as an increase in radicle length.

McDonough (1976) found that kinetin treatment of the

seeds of ten mountain range plants induced earlier germination even at temperatures of 2° C. Puls and Lambeth (1974) found that the initial germination rate of 10-year-old tomato seeds was significantly increased by soaking in kinetin. Makino et al. (1969) found the length and fresh weight of radish was increased after 30 days from sowing by soaking the seeds in 1 ppm of 6-benzylaminopurine or by foliar spray with 4 ppm of 6-benzylaminopurine.

Badizadegan and Carlson (1967) found that treatment of seeds of McIntosh and Wealthy apples with 6-benzylaminopurine at concentrations of 5 to 25 ppm for 24 hours, increases germination by twofold or slightly more.

The Answer

Gibberellic acid is the most potent germination promoter, breaking seed dormancy in a number of crop seed. Lettuce seed will germinate in darkness if treated with gibberellic acid.

Cytokinins have speeded up germination time for many kinds of seed, especially under stress combinations. Although most of the research on an increase in speed of germination has focused on lettuce, the influence of cytokinins is observed on many plants.

Seaweed extracts contain many gibberellins and cytokinins which have been demonstrated to speed up germination.

HOW COULD IT HAPPEN?

Root Growth

Many plant growth regulators have been in use for a long time to initiate, accelerate, and promote root growth of cuttings. Derting found that dips of cutting in seaweed extract solution increased the rooting percentage of **Rhododen-**

dron maximum and **Ilex crenata Rotundifolia.** Featonby-Smith reported that tomato root growth was significantly improved whenever seaweed concentrate was applied, irrespective of whether it was applied as a foliar spray at regular intervals, or whether it was applied to the soil.

Young, working at the University in Melbourne, Australia, found that regular use of seaweed extracts as a foliar spray or as a soil feed encouraged root development in the following crop plants: wheat, sunflowers, beans, corn, peas, and grasses.

The Answer

The most widely researched and widely used plant growth promoter is indolebutyric acid (IBA). Perhaps next in importance is the auxin Naphthalene acetic acid (NAA).

Kingman and Moore reported on two extraction techniques used to isolate, purify, and quantitate several growth promoting substances in seaweed extracts. They reported on the indoles, purines and abscisic acid. Young has reported the presence of growth promoting substances in seaweed extracts. Mowat, as early as 1964, reported the occurrence in algae of two groups of plant hormones, the auxins and gibberellins.

Seaweed extracts contain indole compounds in addition to other plant growth substances that promote root growth.

HOW COULD IT HAPPEN?

Flowering

The ultimate goal in crop production is maturity, whether it is for vegetative parts — Irish potato, sweet potato, lettuce, alfalfa, or for flowers as in floricultural crops — roses, chrysanthemums, carnations, or for seeds such as corn, wheat, beans, peas.

The cumulative effect of all the metabolic changes occurring in growth and development is initiation of flower buds. The vegetative phase of growth determines bud development followed by flower initiation, resulting in the number of flowers produced.

Flowering, or the resultation of it, is the most important practice in agriculture. Some growth regulators are used to promote flowering, some are used to delay flowering, some are used to inhibit flowering.

In 1973 a United Kingdom trial was carried out on three varieties of plum, in which the fate of 1,000 blossoms was followed to maturity. Seaweed extract was applied at a rate equivalent to 11 litres Multiple Concentrate/ha on four separate occasions between mid-April (equivalent to our mid-October) and mid-June (our mid-December). The observations were made on 10 trees from each treatment and the fate of 1,000 blossoms on every tree was followed to harvest. The results are tabulated below:

Variety	Fruit per 100 blossoms (Average of 10 Replicates)	
	Seaweed-treated	Control
Burbank	3.5	2.5
Laxtons Cropper	6.5	3.5
Wyedale	6.8	4.2

It is evident from these figures that seaweed increased the effective fruit set.

The use of seaweed extracts on apricots, cherries, peaches and plums in several Australian states (N.S.W., Victoria, Tasmania) at full bloom has consistently increased the effective fruit set, and has improved the appearance of the fruit. A trial is currently in progress in Victoria on two varieties of peaches to confirm the overseas work regarding the effect of seaweed extracts on the shelf life of the fruit.

Conclusions: Seaweed Multiple Concentrate applied to stone fruit at 6 1/ha at full bloom and with subsequent routine sprays results in greater effective fruit set (i.e., yield), and generally can be expected to significantly increase the shelf life of the fruit. The earlier applications appear to have the greatest effect on fruit shelf life.

Seaweed Multiple Concentrate applied to grapevines at 3 to 4 1/ha (or at a dilution of 1 part seaweed M.C. to 400 parts water) as a foliar spray at bud burst and then with subsequent routine sprays until 12 to 18 1/ha have been applied results in:

(a) increased fruit set, particularly under adverse climatic conditions;
(b) more open bunches, which allows more even development of the fruit and assists in the control of Botrytis.

A French trial to test various foliar nutrients was conducted in 1970 and 1971 at Champagne Mercier by the Interprofessional Committee for Champagne Wines. The cultivar used was Pinot Meunier. The yields in 1971 were about 60% below average, owing to the wet, cold spring — also a hazard in Australian growing areas. The yield, expressed in kg/ha for 1971 were as shown:

Treatment 1970	Treatment 1971	Yield kg/ha in 1971
Control	Control	6328
Control	Control	6770
Seaweed	no seaweed	6963
Seaweed	no seaweed	7200
Seaweed	Seaweed	7337

It is interesting to note that chemical foliar nutrients (NPK plus trace elements) were also included in the trial, and in 1971, these significantly reduced the yield.

Seaweed is used extensively on vines in Europe, the United States, and New Zealand, and in the past few years has become increasingly used in Australia, particularly in the Murray-Goulburn area.

Virtually every grower who has used seaweed on vines in Australia has commented on the fact that seaweed causes the bunches to be more open, enabling the grapes to swell more evenly, and assisting in the control of Botrytis.

Conclusions: Seaweed Multiple Concentrate applied to grapevines at 3 l/ha with routine sprays can increase yield in adverse climatic conditions and results in a more open bunch, which assists in the control of Botrytis.

Seaweed Multiple Concentrate applied to strawberries at 10 to 15 l/ha spread over 4 to 6 applications, commencing at first sign of growth:

(a) increases fruit set;
(b) reduces the incidence of Botrytis cinerea (Strawberry Gray Mould).

The Answer

Gibberellins are known to induce flowering in many plants. It has been reported that growth regulators which promote flowering in one crop could inhibit flowering in a different crop. Critical factors are selection of the correct growth regulator, apply in the recommended amount, apply at the proper time and stage of plant growth.

Plant growth and development, as earlier stated, is controlled by substances (hormones) produced by the plant itself. These hormones regulate or promote normal plant growth. By applying other plant growth regulators the plant made plant growth regulators amount and action are changed, thereby changing normal plant growth.

Increased resistance to fungal diseases by plants treated with seaweed extracts has been widely reported by grower field observations. Many instances are reported in the scientific literature. Senn, et al. (1961), observed that mildew was much less prevalent on melons sprayed with seaweed extract than on the control. Booth (1966) reported reductions in the incidence of **Botrytis** infection in strawberries and a 50% reduction in "damping off" in tomatoes. Stephenson (1966) reviewed research showing that seaweed extracts reduced the incidence of **Botrytis** infection in strawberries significantly.

Why seaweed extracts help plants resist fungal attacks is not fully understood. Treated plants may be nutritionally unsuited to fungal development. Research by Khaleafa, et al. (1975), shows that seaweed extracts have fungicidal properties. Another theory is that the extracts contain products of hydrolysis or digestion from manufacturing processes and these may contribute to reduced fungal attack.

HOW COULD IT HAPPEN?

Cold Hardiness

In 1959 research was initiated at Clemson University to investigate the influence of seaweed extracts on plant growth of certain horticultural crops. The following was reported in 1961.

> *Tomato plants grown in the flats in which seaweed meal had been applied possessed considerable cold resistance. Flats of seaweed treated plants withstood freezing temperatures as low as 29° F while checks were killed. Tomato plants from the various greenhouse seaweed plots were*

> *transplanted in the field during the summer. In late October a heavy frost severely injured the check plants whereas the treated plants survived the first two frosts. More research is needed in this area.*

Booth (1963) reviewed the Clemson report and followed up with this statement.

> *New ideas cause more pain than any of man's other afflictions and writing about them has its hazards. I recorded in the Grower (1963, Vol. 59, p. 1150) some work by Senn in Carolina which claims that liquid seaweed product imparted frost-resistance to tomatoes and citrus fruit. Naturally, this caused me quite a bit of trouble but I had considered his results for two years before writing the article and had sufficient material to support Senn's observations.*

The Answer

Bonner (1943) suggested that plant nutrition might influence growth at the temperatures below the optimum. Some very similar effects on tomatoes have been obtained with nicotine acid by Kettellapper (1963) who advanced the theory that variations from the optimum growth temperature of plants may cause a shortage of one or more essential metabolite. Cathey (1964) reviewed the various effects of growth retardants, especially their effect on the resistance of plants to frost injury.

Numerous investigators since have reported the effects of growth regulators on the ability of plant systems to withstand changes in temperature.

In 1965 Marth stated that frost damage to cabbage plants was significantly reduced by the application of certain plant growth regulators prior to exposure to low temperatures.

In 1966 reports reveal that applications of plant growth regulators produced increases in frost hardiness of winter wheat, tomatoes, and box elder.

In 1976 Rikin stated that abscissic acid was involved in the formation of chilling resistance of cucumber seedlings.

Research reports in 1980 showed that polyamino compounds would protect a wide range of crops from frost damage.

The Tasmanian Department of Agriculture conducted a number of experiments with plant growth regulators in search of a suitable material that would reduce frost damage. A seaweed extract of **DURVILLEA potatorum** produced excellent results.

Garman (1967) of the Connecticut Agricultural Experiment Station reported on research using Norwegian **ASCOPHYLLUM nodosum** on apple trees to improve the number of fruit spurs. He stated, "The trend towards production was first seen here in bearing Red Delicious and Romes when it became evident that even with heavily loaded branches, new fruit spurs produced abundantly alongside the fruit. This data is given in the table below. In these blocks, crops were heavy in 1965 and moderate to heavy in 1966, indicating little or no reduction in off years. General observations on peaches also indicated a heavier set on sprayed as compared with unsprayed trees."

New Shoots Alongside Fruit Counted in June 1964

Treatment	Variety	No. Trees Examined	No. Fruit Spurs Alongside Fruit[1]	Difference
Crow Hill[2]				
ASCOPHYLLUM nodosum	R. Del.	15	42	
Control	R. Del.	15	26	16
ASCOPHYLLUM nodosum	Rome	15	150	
Control	Rome	15	108	42
Hillhouse				
ASCOPHYLLUM nodosum	R. Del.	10	75	
Control	R. Del.	10	13	62

[1]Trees examined by passing completely around each one and making counts of visible shoots alongside apples — usually 3 to 4 inches in length.

[2]Seaweed sprayed on in 5 sprays in Hillhouse block, 6 sprays on Crow Hill.

General observations since 1963 (sprays started in 1961) showed that mite control has been less difficult in the sprayed areas. Similar results were seen in other Connecticut orchards.

There is, however, no evidence that the seaweed materials supply all the nutrients needed to maintain orchards in full vigor through reduction in the total amounts applied.

So far we have seen no evidence of incompatibility with pesticides commonly used in Connecticut.

Hormones, plant growth regulators, are not new, but scientists are just beginning to explore and develop their great potential in agriculture.

HOW COULD IT HAPPEN?

Insects

As early as the 1940's, there were reports that plant growth regulators could be valuable in controlling certain insect pests. In 1960 Senn observed reduction in aphid and flea-beetle populations on plants that had been sprayed with seaweed extracts. In 1964 Driggers and Marucci reported on the reduction of European red mite populations on peaches and aphids on strawberries where extracts of **ASCOPHYLLUM nodosum** were used as a foliar spray. Mite counts were made weekly throughout the season and showed 40% fewer adult mites on the seaweed extract treated plots than on the check plots.

Adult Mites Per 100 Leaves

Check					Seaweed				
1964	1965	1966	1967	1968	1964	1965	1966	1967	1968
69	974	2	0	72	80	802	3	0	77
12	700	8	0	163	14	282	15	0	275
1	488	0	2	—	1	60	1	3	—
13	2662	4	0	401	11	502	3	0	244
5	184	2	2	675	0	24	1	1	628
5	3578	38	0	1157	2	92	33	2	711
127	2648	30	5	897	7	54	17	2	419
194	2918	93	17	453	111	800	67	22	267
327	22	299	52	174	104	34	198	38	72
559	106	265	42	72	208	554	117	77	18
814	46	372	269	22	218	196	369	198	2
1062	110	809	386	5	453	782	366	257	2
3188	14436	1922	775	4091	1209	4182	1190	600	[a]2715
265	1203	160	65	372	101	348	99	50	[b]247
					62	71	38	23	[c]37

[a]Totals
[b]Average per collection
[c]% reduction over check

In 1969, Driggers reported on a five-year study on the European red mite population reduction on Red Delicious apple trees when seaweed extracts had been used.

Driggers concluded that the Norwegian seaweed extract applied as a foliar spray on apples appears to suppress the mites high capacity for reproduction. Whether this occurs directly or indirectly is not evident. Also what compound or compounds present in the seaweed extract that brings about the observed effect is not known. It may be some hormone or auxin that is absorbed through the leaves of the apple tree. Whether all seaweed extracts have this property or is limited to the **ASCOPHYLLUM nodosum** species is yet unexplored.

Austin, et al. (1965), reporting on apple research from England stated that, "Clearly, the treatment with Norwegian **ASCOPHYLLUM nodosum** held down the red spider mite population to very low levels." **ASCOPHYLLUM nodosum** contains chelated metals, and experiments by Terriere (1964) have shown that the use of chelated metals leads to greatly reduced fecundity in red spider mites.

The Answer

Eichmeier reported that two-spotted spider mites raised on snap beans showed that treatment with gibberellic acid resulted in significant reductions in mite populations. Rodriguez reported similar effects on gibberellin treated bean and apple trees.

Hormones are reported to keep certain insects sexually immature so they can't reproduce. Hormones have also been reported to prevent female insects from producing sex attractants they need in order to mate.

Booth sums the situation up by saying:

Taking the situation as a whole, there is experimental evidence to show that liquid seaweed products, trace elements, and growth retardants have an adverse effect on the reproduction of insects. There are also practical experiences with seaweed products and suggestions, from spraying contractors that a mixture of liquified seaweed and an insecticide is more efficacious than the insecticide alone.

HOW COULD IT HAPPEN?

Nematodes

In the 1960's researchers at Clemson University observed that tomato roots and okra roots grown in seaweed extract (**ASCOPHYLLUM nodosum**) treated plots had fewer nematode infestations than comparable roots from the control plots.

Featonby-Smith and Van Staden (1982) stated that seaweed concentrate (**ECKLONIA maxima**) at a dilution of 1:500 improved the growth of tomato plants significantly, irrespective of whether or not it was applied as a foliar spray at regular intervals or whether the soil medium was flushed once with diluted seaweed concentrate at transplanting. They commented further that it was significant that root growth was much improved whenever seaweed concentrate was applied and, in particular, that root-knot nematode infestation was visibly reduced in all cases where the seaweed concentrate was applied. This undoubtedly resulted in improved root development, and thus more efficient moisture and nutrient utilization by the plants. Results show that although the number of nematodes increased in the soil after the application of seaweed concentrate, the numbers which had established themselves in the roots was reduced when compared to this control.

Paracer et al. (1985) evaluated aqueous extracts of 12 Florida marine algae for control of various species of nematodes. They reported that **Spatoglossum schroederi** was effective in killing **Meloidogyne incognita, M. javanica, M. acrita** and **Hoplolamus galeayus** at concentrations of 1.0%, 0.75% and 0.50%. They also stated that tomato plant growth was significantly improved by these algae.

The Answer

Seaweeds contain antibiotics including bromophenols, tanning, phloroblucinol and terpenoids. Dropkin (1969), Kochba (1971), Sawhney (1975) all report on the role of hormones, especially cytokinins, in nematode infestation and development in the roots of susceptible hosts.

Earlier chapters of this publication have listed numerous listings of reports documenting the presence of cytokinins in various seaweed extracts.

Dropkin (1969) has shown that high concentrations of kinetin inhibited both larval penetration and development in the roots of tomato plants. The levels of cytokinin present in seaweed concentrate has a similar effect on nematode infestation and development.

Brueske (1972) stated that infestation of the roots with **Meloidogyne incognita** resulted in decreased cytokinin levels in the root exudate of tomatoes.

Featonby (1982) concludes that this decrease in cytokinin translocation to the shoots may be responsible for the reduced shoot growth associated with nematode infestation. The application of cytokinins found in seaweed concentrate may be instrumental in overcoming this imbalance.

Tarjan (1973-80-83-84-85), Professor and Nematologist, University of Florida, Gainesville, Florida, has published many reports on his research using seaweed products. In a recent report he mentions that the treatments with proprietary nematicides, and with Norwegian **ASCOPHYLLUM nodosum** @ 28 liters 1 ha, appeared to promote a stronger root system.

HOW COULD IT HAPPEN?
Shelf Life

A test was conducted in 1963 in an orchard of late Rio-Oso-Gem in Gaffney, South Carolina. Peaches were sprayed with seaweed before harvest and comparable lots of untreated fruit were held in the laboratory and examined daily until 3 weeks after their harvest (Table 1). At the end of this period, 3 times as many untreated fruit had deteriorated as treated fruit. Thus, the same trend in improved shelf life as a result of the seaweed treatments was found as in the fruit from the Sandhill Station.

The treated fruit were slightly more firm than the control at harvest, although the differences were so small the extended shelf life of the seaweed treated fruit could not be explained on the basis of firmness alone.

Only slight differences were noted in pH, total acidity, soluble solids, ground color, puree color and moisture content between the treated and untreated lots of fruit.

In 1964 an experiment was conducted using the same orchard of late Rio-Oso-Gem variety as used in 1963. One exception in the 1964 experiment was that not only seaweed sprays were used, but also seaweed meal at a rate of 2 lbs/tree for 4-year-old trees. The seaweed meal was applied to the soil in January. The sprays were to begin 1 week after harvest and be carried on every 4 weeks up until 1 week before harvest. This experiment was set back due to a late spring freeze which killed the entire Rio-Oso-Gem variety.

The following year the experiment was again conducted

in the same orchard. Trees receiving seaweed sprays previously and ones with no previous application of seaweed were selected. These trees were located in different sections of the orchard, and were treated as two blocks. Each block then received ten treatments which consisted of application of seaweed meal with varying numbers of seaweed spray. Treatments are shown in Tables 2 and 3. Fruit were harvested and shelf life studies were conducted.

As shown in Tables 2 and 3, there appear to be a beneficial residual effect of seaweed sprays. Generally, percentage of peaches rotten from plots having previous applications was lower than those having no previous sprays. Seaweed meal applied to the soil resulted in approximately a 50% decrease in rotten peaches. The decrease in rotten fruit from trees having no previous sprays was lower than this, decreasing from 43% to 24%. This is in agreement with the work of Childers and Driggers.

Table 1

Shelf Life of Rio-Oso-Gem Peaches
Harvested September 5, 1963

Treatment	Number of Days After Harvest													
	9	10	11	12	13	14	15	16	17	18	19	20	21	Total
Check	Number Rotten:													
1	6	2	1	0	0	2	0	4	0	0	1	1	0	17
2	2	5	1	0	0	0	0	3	0	1	3	0	2	17
3	3	5	1	0	0	0	0	2	0	0	3	2	1	17
Total	11	12	3	0	0	2	0	9	0	1	7	3	3	51
Treated														
1	0	0	1	1	0	0	0	0	0	1	0	1	1	5
2	0	0	0	0	0	1	0	1	0	1	0	0	0	3
3	3	0	0	0	0	0	0	0	1	1	0	1	3	9
Total	3	0	1	1	0	1	0	1	1	3	0	2	4	17

Table 2

Effect of Various Combinations of Seaweed Meal and Extract on Shelf Life of Harvest Gold Peaches, Gaffney, South Carolina, 1965*

Treatment**	Percent Peaches Rotted Days After Harvest			
	7	14	21	28
Check***	5.6	13.3	20.0	43.3
Check + Meal	1.7	7.2	12.8	23.9
2 Sprays	4.4	8.9	17.8	28.9
2 Sprays + Meal	2.8	9.4	14.4	34.4
3 Sprays	1.7	5.6	10.6	25.6
3 Sprays + Meal	5.0	14.4	27.2	43.9
4 Sprays	4.4	7.2	11.1	31.7
4 Sprays + Meal	4.4	13.9	21.2	31.7
5 Sprays	3.9	7.2	13.3	29.4
5 Sprays + Meal	7.2	17.8	28.3	46.1

*These trees had no previous seaweed treatment.

**Where meal was used, 2 pounds were applied in the fall to each tree. Concentration of spray material was 1 gallon of seaweed extract per 100 gallons of water applied until drip-off.

***Regular spray schedule.

Table 3

Effect of Various Combinations of Seaweed Meal and Extract on Shelf Life of Harvest Gold Peaches, Gaffney, South Carolina, 1965*

Treatment**	Percent Peaches Rotted Days After Harvest			
	7	14	21	28
Check***	3.9	7.8	15.6	30.6
Meal Check	2.0	9.4	15.0	20.6
2 Sprays	3.9	9.4	13.3	23.9
2 Sprays + Meal	4.4	13.3	23.9	36.7
3 Sprays	9.4	15.6	20.6	28.1
3 Sprays + Meal	1.1	8.9	13.3	21.1
4 Sprays	5.6	8.3	11.7	23.9
4 Sprays + Meal	5.6	8.9	10.6	26.7
5 Sprays	5.6	9.4	12.8	26.1
5 Sprays + Meal	2.2	5.0	10.0	22.2

*This test plot had seaweed extract sprays the previous year.

**Where meal was used, 2 pounds were applied in the fall to each tree. Concentration of spray material was 1 gallon of seaweed extract per 100 gallons of water applied until drip-off.

***Regular spray schedule.

The Answer

Senescence — Old Age — Death.

Senescence is the deterioration of cells, tissues, organs that ends the functional life of an organism. Senescence is not limited to fruit as it occurs in leaves, flowers, stems and roots.

There are many suggested reasons as to what causes senescence in plant parts. Senescence may be the result of an altered hormone balance in the plant that is associated with the enzyme system. Senescence may be the final differentiation (change) that occurs from germination to death in plants. Some researchers state that senescence results from a hormone substance that is translocated from the fruit to the leaves and other vegetative parts that result in death.

The objective in some phases of crop production is to delay or inhibit senescence. There are many reports that conclude that cytokinins are effective senescence retarding growth regulators.

It is reported that NAA, auxins, and gibberellins will delay senescence in certain plant organs. These growth regulators are widely used in commercial fruit marketing. Seaweed extracts are known to delay senescence in plant organs.

Blunden, et al. (1978), reported that "The shelf life of capsicums was significantly increased by immersion in seaweed extracts." The most noticeable effect was with limes, the rate of "degreening" of fruits after immersion in either seaweed extracts or kinetin solutions being reduced. They state that this marked reduction of the rate of "degreening" of limes produced by immersion in seaweed extracts or kinetin solutions is of potential economic value.

A Survey of Grower Usage of Seaweed and Seaweed Products

The purpose of this section is to present the use of seaweed and seaweed products in agriculture in the United States, and growers' reactions to the products. The information herein does not cover every geographic area where seaweed products are used. However, the information is representative of the growers and their geographic locations throughout the United States where data were collected.

The actual "farm use" of seaweed (liquid and meal) is relatively new in some areas, while in others growers have been using seaweed in food and fiber production for many years. With increasing interest in alternative methods of crop production, this survey was an effort to determine the role of seaweed products in such programs. Funds for this study were made available through the Department of Commerce, Economic Development Administration. The research team from the Department of Horticulture of Clemson University visited a number of farms throughout the United States in order to better understand "farm use" of seaweed. Manufacturers of several seaweed products provided lists of farmers who have used their products for visitation and accumulation of data.

Endorsement of any specie or product to the exclusion of any other specie or product of alternate species is not to be implied.

One of the more popular nurseries in Orlando, Florida, has been using **ASCOPHYLLUM nodosum** foliar sprays for 12 years. There was a period of one year in which the establishment did not use **ASCOPHYLLUM nodosum.** During this time the grower said that both he and his customers felt that his plants did not look as healthy as in past years. "I don't know if you can attribute this to the foliar sprays, but I've been using it every year since and have not had any trouble. I use **ASCOPHYLLUM nodosum** on every plant that goes through this nursery." When asked why he used seaweed he replied, "I am in this business to make money, and when people come here and purchase a healthy plant, they are happy. When they are happy they will return here for their next purchase." When pressed further, he said, "I feel seaweed **(ASCOPHYLLUM nodosum)** contains certain growth hormones and trace elements necessary for proper growth. The price of this product allows me to be sure these necessary things are supplied."

An avocado grower in Homestead, Alabama, has noticed some very interesting things during the 10 years he has been strictly an "organic" grower. "During the first five years of operation we nearly went out of business. Insects and disease nearly took us away. However, from the first year, things progressively got better. We didn't notice disease or certain insect damage to be as severe the fourth year as it was the preceding years. Today we have very, very little problems with insects or disease. I don't know that the use of seaweed **(ASCOPHYLLUM nodosum)** has been responsible for this, but we use seaweed foliar spray on our crops every year." This grower felt the use of organic materials, including seaweed, is often a long range management program and that the real benefit comes after several years of use.

In Labelle, Florida, a grower has been using **ASCOPHYLLUM nodosum** foliar sprays for 12 years. During this time he has made unusual observations. He farms 400 acres of various vegetables, primarily peppers. His acreage is a sandy loam and relatively poor soil. When he first began using seaweed, he says that he really didn't believe that the seaweed was helping him. However, "Over the years I've been able to detect some of the benefits. I am the only farmer who uses seaweed in this area that I know of. Not too many years ago we had a late frost and I was about to get into my truck to see how much damage I had. A neighbor, who has peppers in a field next to mine, drove into my yard and told me there was no need to go to my field. He said he had just returned from his field and everything was ruined. We both got into my truck and went to my field anyway. To our amazement none of my plants was damaged. I had made my first **ASCOPHYLLUM nodosum** application the day before." He also says that he doesn't have the insect and disease problem many of his neighbors have. Two applications of seaweed are made during the growing season with ground rigs. One gallon of seaweed with 40 gallons of water per acre with 40 psi (nozzle) is sprayed each application.

"Time of application seems to be very critical. Our first application is at the 3-leaf stage and the second just prior to bloom."

In Onarga, Illinois, a farmer has used **ASCOPHYLLUM nodosum** foliar spray and meal on oats, alfalfa, corn, and soybeans for about 15 years. He heard about the product through one of his neighbors and uses it because he gets fewer insect and disease problems. He also gets higher yields. "My cows are much healthier by using free choice meal and I believe the use of seaweed has improved my breeding program for dairy cows."

A seed treatment is used on corn and soybeans using 1¼ lbs. of seaweed concentrate with 30 gallons of water. Seeds are not soaked, but rather a fine mist is used to coat the seeds. "The condition of my soil is poor when compared to soils in nearby locations. However, the use of **ASCOPHYLLUM nodosum** has improved my farming efficiency."

A farmer in Evansville, Illinois, has used foliar sprays of **ASCOPHYLLUM nodosum** on 20 acres of corn, 50 acres of soybeans, and 50 acres of alfalfa for five years. "Seaweed gives me a faster recovery rate after cutting alfalfa, better yield, less insect damage, and I get a premium for alfalfa hay treated with **ASCOPHYLLUM nodosum** when sold to race horse stables." The premium price of alfalfa is because of the high protein content. Soybean seed is treated prior to planting using ½ pint of powder/1½ bushel of seed. A foliar application of 1 gallon seaweed concentrate with 10 gallons water/A is applied by air on alfalfa after each cutting.

On a race horse farm in Sparta, Illinois, **ASCOPHYLLUM nodosum** foliar sprays have been applied on 50 acres of alfalfa and 25 acres of additional pasture land for grazing. This farmer had observed that his race horses preferred grazing on seaweed treated strips when he first began using the product. He has found that his hay cures better, is higher in protein, and has better keeping quality than the untreated areas. "Everything I grow on this farm now receives **ASCOPHYLLUM nodosum** foliar sprays (1 gal./10 gal. water/A; air application). My horses are winning races with 5 first place winners in 1977 (2-year-olds; 3-year-olds). All of my management practices are contributing to the success, including seaweed. I am winning and don't intend changing any of these practices."

On another farm in Sparta, Illinois, seaweed foliar sprays have been applied on 1,000 acres of alfalfa and 800 acres of soybean. Quick recovery and higher protein has been observed over the past five years of use. Additionally, the grower has noticed fewer problems with alfalfa weevil. Noticeable results on soybeans haven't been as consistent. "I feel seaweed has helped my soybeans 2 out of the 5 years I've been using it. With 2 very good years, I'm still ahead."

Seaweed has been used for 10 years on a farm in Greenwood, Indiana. The crops now being produced are 200 acres of soybeans, 200 acres of alfalfa, and 25 acres of vegetables. "I used to have treated and untreated areas but not anymore. I use **ASCOPHYLLUM nodosum** on everything." Higher yields and improved quality have been obtained on this farm with the addition of seaweed sprays to the program. Soybeans are sprayed (1 gallon seaweed/10 gallons water/A) prior to bloom. The tomatoes get two applications; once prior to bloom and after first fruit set. Alfalfa is sprayed after each cutting.

Similar results have been observed on 200 acres of soybean and 200 acres of alfalfa in Kokomo, Indiana. Seaweed has been used on this farm for five years with consistent returns.

Fifty acres of seaweed treated alfalfa in Chester, Illinois, have produced higher yields and better quality. "My neighbors and I have been testing this product for the past two years. Based on what we have seen, we will be increasing our acreage progressively in future years. It is important for us to use products that won't pollute."

A nursery in Homewood, Illinois, has been testing **ASCOPHYLLUM nodosum** for two years on woody ornamentals, foliage plants, and fine turf. The grower also sells seaweed products through a supply house. Foliage plants receive sprays once every two weeks while turf and woody ornamentals were sprayed once every two weeks during the growing season only. In this case, applications of **ASCOPHYLLUM nodosum** were made in conjunction with the regular spray program.

As previously stated, this nursery also sells seaweed through a supply house. "We have many repeat customers who have been using foliar sprays in their homes and gardens for several years. These people seem to be happy with the product. I've never had a customer to be dissatisfied with seaweed."

Another nursery in Indianapolis, Indiana, has been using and selling seaweed for six years. The application rate is 1 quart of **ASCOPHYLLUM nodosum** liquid to 50 gallons of water blended with Peters 20-20-20, applied once every two weeks. Crops are primarily house plants, bedding plants, African violets, and woody ornamentals. The nurseryman states that his foliage plants have a better appearance and his woody ornamentals are healthier. "The nursery industry is a very competitive business in this area, and I need anything to give me a slight advantage. I have many repeat customers, and I feel that **ASCOPHYLLUM nodosum** seaweed foliar sprays have helped me."

In Tennessee, Kentucky, Mississippi, Louisiana and Alabama, **ASCOPHYLLUM nodosum** plus fish is being used on a number of crops, i.e., corn, soybeans, cotton, tobacco, vegetables, and forage crops. Use of the product is reported with varying degrees of success. Some farmers report significant increases in yields while others question its economical feasibility. No one reported detrimental effects from the foliar sprays.

A soybean producer in eastern Tennessee observed that his treated soybeans appeared to withstand the drought much better than the untreated soybeans. This was also observed by a corn producer in western Tennessee. This dry period caused problems other than the drought. Some plant diseases seem to flourish in hot, dry weather and this is especially critical when you consider that the plant is already in a weakened and susceptible condition. These farmers felt that the foliar sprays of **ASCOPHYLLUM nodosum** and fish may have been enough to provide their crops with some disease protection. "I don't know that the sprays actually controlled the disease or that the plants were healthier and had more resistance. I do know that my treated soybeans were healthy during the drought."

These farmers used 1 pt/A **ASCOPHYLLUM nodosum** plus fish foliar spray with 10 gallons of water and applied the material by air. A "one-time" application was all that was necessary and in most cases this was applied before flowering.

The same product has been used extensively in the midwest primarily on corn, soybeans, and alfalfa. Says one northeastern Iowa corn producer, "I've been farming in this area for many years and oftentimes during adverse conditions. Many of my neighbors laughed when I first began spraying seaweed plus fish on my corn. Today, many of these same neighbors are using this product because we have compared treated and untreated fields. We all agree that the investment is small ($10.10/A) and returns are consistently higher." In this particular area of Iowa alone, there were 60,000 acres of treated corn. Soybean producers in the same area have observed similar results. Alfalfa producers in this area agree that their livestock prefer to graze on strips that have been treated with seaweed plus fish. One grower had his alfalfa tested for protein content and found that the treated alfalfa was higher in protein. He says, "This is the reason my livestock prefer it, and this is the reason I will continue to use it."

In southeastern Minnesota a potato producer had the best year he ever had. "I've farmed this land for 15 years and I am convinced seaweed plus fish boosted my yields this year. We were very busy and didn't have the time to leave untreated strips, but I still feel I've seen enough to continue using the product." This particular area in Minnesota is very high in organic matter. One of the major problems of food production is a deficiency of minor elements which is typical of most muck-type soils. It is highly possible that the seaweed plus fish has supplied these minute quantities of trace elements already proven to be essential in food and fiber production.

Some farmers in this midwestern area have experimented with soil applications of liquid seaweed plus fish at the time of planting as a replacement for starter fertilizer on soybeans. The material is sprayed on top of the soil in a 3- or 4-inch band over each row at the time of planting. Other farmers have cut their starter fertilizer by 25% on corn and have sprayed a soil application of seaweed plus fish. One quart of seaweed plus fish with 8-10 gallons of water per acre seemed to be the most used application as a starter. Later, usually before fruit set, an additional quart of seaweed plus fish per acre is applied by air.

A camellia grower from Fayetteville, North Carolina, U.S.A., reporting in the *Camellia Magazine*, Vol. 22, No. 5, Nov. 1967, stated, "I don't believe I have ever seen plants with a darker, prettier green color than ours had. We decided that indeed the seaweed was an excellent source of trace elements if plants could be that healthy in our poor soil. Our blooms were 10%-15% larger and the life of the blooms was lengthened 2-3 days."

Conclusions

It appears that the use of seaweed and seaweed products is growing throughout the United States with the most usage being in the southeast and midwest. Seaweeds are used in the west primarily in formulations with other materials. However, many western farmers use seaweed directly on their crops in the form of foliar sprays.

It is still not known what impact seaweed will have on food and fiber production in the future. Farmers who have been using seaweed successfully feel that seaweed can be used to advantage in their programs. Not all farmers have observed improved yields. However, it appears that the majority of these farmers surveyed herein have seen improvements to justify further investigations and "on-farm" testing as well as the continued use of seaweed in their crop production.

Points to Ponder

Dr. Metting of R and A Plant-Soil, Pasco, Washington (1985), conducted evaluation tests of **ASCOPHYLLUM nodosum** liquid extract as a foliar spray for dryland winter wheat **(Triticum aestivum** L. Var. "Stephens") in the inland Pacific northwest. He reported dramatic differences when the seaweed was used as a foliar spray. Treated plots averaged better than 14 bushels per acre greater yield than the control plots. The data for number of culms per plant suggest that the positive influence on yield probably resulted as a consequence of the ability of the seaweed foliar spray to minimize herbicide damage. Brian, et al. (1977), reported on the enhancement of herbicidal effect by seaweed extracts in marine natural products chemistry.

Crawford (1964) reported on sweet potato. Field plots of Centennial sweet potatoes were set up in 1962 at Clemson University using varying rates of seaweed meal in comparison with additional organic materials. Results indicate increased yields of No. 1 and 2 sweet potatoes from seaweed treatments. Additional organic acids tended to depress yields. Soluble solids were increased with applications of seaweed meal.

Field tests were conducted in 1963 using Caro Gold sweet potatoes to compare two rates of seaweed meal and two rates of 5-10-10 fertilizer. The addition of seaweed meal alone increased the yields. However, when seaweed meal was applied along with 5-10-10 fertilizer, the increase was approximately twice that of the seaweed meal alone.

The Influence of 100 ppm Weekly Sprays of ASCOPHYLLUM nodosum on Nodulation of Soybean
Simpson Farm, 1974, S.C.

Treatment	Mean Count
Control	96.0 a*
Seaweed Spray	102.5 a

*Means separation by Duncan's multiple range test at 5%.

The Influence of 100 ppm Sprays of ASCOPHYLLUM nodosum on Yield of Okra
Simpson Farm, 1974, S.C.

Treatment	Weight (g)	No. Pods	Size
Control	7483 a*	412 a	18.16 a
Seaweed	9469 a	487 a	19.61 a

*Means separation by Duncan's multiple range test at 5%.

The Influence of 100 ppm Weekly Sprays of ASCOPHYLLUM nodosum on Fiber Content of Okra
Simpson Farm, 1974, S.C.

Treatment	Fiber Content %
Control	4.68 a*
Seaweed Spray	4.70 a

*Means separation by Duncan's multiple range test at 5%.

Brian, et al., investigating the enhancement of herbicidal effect of seaweed extracts, reported that "The results indicate that a combination of seaweed extract with a herbicide can have beneficial effects in terms of both weed kill and protection of the crop damage by the herbicide, and it is possible that this may have commercial significance, although it is not known whether the phenomenon is restricted to auxin-type herbicide or even to CMPP alone.

Summary

Those who do not read of the advancements taking place in his own field of endeavor have no advantage over those who cannot read.

There is experimental evidence and there is practical experience to show that seaweed extracts have beneficial effect on plant growth and development.

Many plant processes can be changed by the use of plant growth regulators. The use of plant growth regulators will increase dramatically in the near future. Many scientists are making progress in the basic understanding of the biochemistry of "how plants grow."

You can choose to believe it or write it off as just so much "hogwash." In all such instances of maybe-maybe-not, it is up to the individual to weigh the facts, consider the possibilities, and draw a personal conclusion.

Remember that past generations had their opinions. Mythological gods? A spherical earth? Electricity? The war to end all wars? Television? Computers? Missiles? Men on the moon? Still, who's to say?

Go ahead. Form an opinion. But after you take your position, keep an open mind. Because, if you don't, what you proclaim to be "hogwash" today may be proven to be fact tomorrow. Read and ask questions, but do not "condemn what you do not understand."

Scientists have long since proved that there is no discernible limit to the number of bits of information that the human mind can receive, absorb, understand, and remember.

A wise grower will hear and increase learning; and a man of understanding shall attain unto wise counsels.

Effect of increasing concentrations of plant growth regulators in **ASCOPHYLLUM nodosum** on the flowering of geranium.

Selected References

1. Austin, M.D. 1965. Grower. No. 3600, 29.
2. Badizadegan, M. and R.F. Carlson. 1967. Proc. Amer. Soc. Hort. Sci. **91** 1.
3. Biddington, N.L. and T.H. Thomas. 1976. Antibiotic (fungicidal) action from extracts of some seaweeds. Botanica Marina 18. 163-5.
4. Booth, C.D. 1964. Grower. **62,** 442.
5. Booth, E. 1966. Some properties of seaweed manures. Proc. 5th Int. Seaweed Symp., pp. 349-57. Pergamon Press: London.
6. Cathey, H.M. 1964. Ann. Rev. Plant Physiol. **15,** 271.
7. Cook, R.L. 1983. The use of seaweed extract **(ASCOPHYLLUM nodosum)** for pregermination of onion sets. Hort. 457. Mich. State Univ., East Lansing, Mich.
8. Driggers, B.F. and P.E. Marucci. 1964. Hort. News. Rutgers Univ.
9. Dropkin, V.H., et al., 1969. J. Nematol., **1,** 55.
10. Esashl, T., et al., 1975. Aust. J. Plant Physiol. **2,** 569.
11. Featonby-Smith, B.C. 1982. Sci. Hortic. (AMST) **20** (2), 137.
12. Garman, P. 1967. Conn. Agric. Expt. Sta.
13. Khaleafa, A.F., et al., 1975. Antibiotic (fungicidal) action from extracts of some seaweeds. Botanica Marina 18. 163-5.
14. Kingman, A.R. and J. Moore. 1982. Botanica Marina. **25** (4), 149-154.
15. Kochba, J. 1971. J. Amer. Soc. Hort. Sci., **96,** 458.
16. Kuribayashi, T. and H.O. Ohashi. 1975. Shavakugaku Zaashi **29,** 62.
17. McDonough, W. 1976. Phyton. (Buenos Aires), **36,** 41.
18. Makino, H., et al., 1969. Tamagawa Daigaku Nooakubu Kenkyu Hokoku, **9,** 78.
19. Marth, P.C. 1965. Jour. Agric. Food Chem., **13,** 331.

20. Puls, E.E., Jr., and V.N. Lambeth. 1974. J. Amer. Soc. Hort. Sci., **99,** 9.
21. Reynolds, T. and P.A. Thompson. 1973. Physiol. Plant 1973, **28,** 516.
22. Rikin, A. 1976. Physiol. Plant, **38,** 95.
23. Sawhney, R. 1975. Nematologica, **21,** 95.
24. Sharples, G.C. 1973. J. Amer. Soc. Hort. Sci., **98,** 209.
25. Tarjan, A.C. 1983. Nematropica. **13,** 1, 55-62.
26. Young, H. 1977. Annal. Biochem., 79, 226.

Effect of varying rates of seaweed meal in soil mixtures on growth rates of geranium.

CHAPTER 11
Speak the Language

Man knows much more than he understands (A. Adler, Social Interest).

Imagine visiting in a foreign country and being with a group of people talking, but you don't understand a word you hear. This situation could be repeated on a farm when a dealer talks biological farming. Therefore, it is important that dealers, salesmen, and growers "speak the language" of today's biological agriculture.

Bio-Living

Truly, growers are biologists, for they are dealing with the science of life; the branch of knowledge which deals with living organisms and vital processes. Biological farming deals with life and living processes. The soil is teeming with life, or it should be, if growers are to produce satisfactory yields. This is where marine algae-seaweed fits into the big picture of farming.

Bio-Chemical

This is the chemistry that deals with the processes occurring in living plants. Photosynthesis, the most vital of all biochemical processes, is using the energy of the sun to manufacture food for plants, animals, and man. Another biochemical process, sometimes not appreciated, is respiration, the process of releasing energy from manufactured foods.

Bio-Physical

This is the physics dealing with processes occurring in living plants. Transpiration is a biophysical process by which plants give off water through openings in the leaves. Water absorption is another essential biophysical process. Water is absorbed in the liquid state in the regions of cell elongation and the root-hair zone, and water is lost in the vapor state from the tissues of plants. Without these processes the manufacturing of foods by plants would stop.

It becomes evident, after understanding the "bio-language," how vital the products made from marine algae-seaweed are to successful farming operations.

Essential Elements

The essential elements required by crop plants are called plant nutrients. The major or macronutrients necessary for plant growth and development are carbon, oxygen, hydrogen, nitrogen, phosphorus, potassium, sulfur, calcium and magnesium.

The micronutrients are also essential elements but are required by plants in smaller amounts. The micronutrients necessary for plant growth and development are manganese, iron, boron, zinc, copper, molybdenum, and chlorine.

Plant growth in all its phases is dependent on the micronutrients; thus, the importance of seaweed in biological farming.

Another group of elements that are necessary for growth and development of certain specific plants are called functional elements. These elements are sodium, cobalt, silicon, vanadium, and possibly aluminum and selenium. It is reported that silicon is essential for the growth of tomatoes, the amounts needed are less than 0.2 micrograms per gram of dry plant tissue. Certain desert plant species both in Australia and the United States require sodium. Silicon is reported to stimulate the growth of members of the grass family as well as certain marine algae. Cobalt is also necessary for the growth of blue-green algae, and it is required by certain legumes to fix atmospheric nitrogen. Vanadium is required by certain green algae.

There is much to be learned about plant nutrition, and especially as it relates to conversion to amino acids and finally to human and animal nutrition.

Chelated Micronutrients

The word **chelate** is derived from a Greek word meaning claw. The Greek word certainly fits because chelates are large organic chemical structures which encircle and tightly hold micronutrients iron, manganese, zinc and copper.

Chelating agents were first used in nutrition studies about 1951. Chelated micronutrients are now commonly used to correct deficiencies of iron, zinc, manganese, and copper. Plant roots can absorb the micronutrient-chelate combination. The micronutrient can then be released inside the plant.

The high solubility of chelated micronutrients make them very desirable to fertilizer formulators.

It has been known for many years that naturally occurring chelates of micronutrient cations exist in the soil. It is also known that certain proteins, amino acids, flavonoids, organic acids and purines present in organic matter are capable of chelating micronutrients .

Seaweed is a source of micronutrients such as iron and chelating compounds such as mannitol (refer to chapter 2).

The amounts of micronutrients required by plants will naturally vary somewhat with the particular plants and the annual micronutrients requirement will obviously vary with how much of the crop is removed and how much micronutrients are recycled. When seaweed extracts are used at the recommended times and rates it will supply the amounts of iron, zinc, copper, molybdenum, cobalt, boron, manganese and magnesium that most crops require. Seaweed extracts contain chelating compounds which will make some micronutrients already in the soil (but which are not available to plants because of the solubility) available to the plants in a chelated form. While there are many sugars in seaweed which could chelate, the major chelating agent is mannitol. Chelated elements may be applied to the soil or as foliar sprays. The micronutrients in seaweed are in chelated form and seaweed containing mannitol aids in additional chelation.

Seaweed Contains Growth Regulators

Plant growth and development is controlled by chemicals produced by the plant itself called hormones. Plant hormones are naturally occurring plant substances including auxins, gibberellins, and cytokinins. Auxins and gibberellins are compounds that promote cell enlargement and cell elongation. Cytokinins promote cell division and cell wall formation in plants. There are many other physiological effects of cytokinins on plant activity.

The micronutrient zinc activates the enzyme system that promotes the production of auxin. A deficiency of zinc prevents the terminal growth of pecan and peach limbs resulting in a situation called rosette. This deficiency may be corrected by soil amendment or a foliar spray of zinc sulfate. The immediate result is an increase in auxin and a renewed growth of terminal limbs.

Seaweed contains the growth regulators which are collectively termed cytokinins. Research using bioassays and chemical analyses in addition to gas chromatographic analysis support this conclusion.

The most important constituent of seaweed extracts are the plant growth regulators, but not all the value of seaweed extract based products can be linked to the identified plant growth regulators.

The plant growth regulators which have been shown by researchers to be present in seaweed include auxins, gibberellins, and cytokinins (adenine, purines, kinetin, and zeatin). Probably the most important of these plant growth regulators in seaweed are the cytokinins. More research has been conducted with these regulators than the others.

Cytokinins delay senescence in a wide variety of plants. Not only do cytokinins delay senescence in plant leaves but have been shown to be effective in delaying senescence in petals of various flowers. These delays in senescence will result in more active leaves and flowers resulting in higher yields.

Plant Foods

The manufactured products of plants are called plant foods and consist of carbohydrates, sugars and starches, proteins, and fats and oils. All of these are made up of elements assimilated by the plant and supply the food needs of the world.

Parts Per Million (ppm) — Parts Per Billion (ppb)

The successful grower must learn to speak in terms of PGRs (Plant Growth Regulators). These substances are present in plants and those commercially available are also used by the grower to regulate plant growth. They are used in extremely small quantities so the grower must be familiar with ppm (parts per million) and ppb (parts per billion).

Dr. Fred Davis of the University of Florida states that a ppb is equivalent to having a single red grapefruit in the midst of 25,000 trailerloads of white grapefruit.

It is therefore understandable why some people can't understand how such a small amount can accomplish so much. A look at modern-day medication will help to appreciate how such small amounts of a drug can control pain, reduce blood pressure and aid in body conditioning. These drugs are issued in milligrams or lower.

Growers must learn to calculate ppm and ppb for concentration of plant growth regulators, herbicides, seaweed extracts and fertilizers.

Dr. John W. Kelly, Clemson Horticulturist in Palmetto Plant Talk, gives the following methods for calculating parts per million.

Calculating Parts Per Million

Greenhouse growers frequently express the concentration of fertilizers, in terms of parts per million (ppm). This unit of measure is relatively unique to the greenhouse industry and often there is some confusion on how ppm is calculated. The following is a "simplified" formula suitable for most greenhouse applications.

I. To calculate the ppm contained in 1 ounce of material, first solve for B:

$$A \times 75 = B$$

Where:
A = the % active ingredient (AI) in the fertilizer
B = ppm contained in 1 ounce of the material in 100 gallons of water.

Example: Calcium nitrate contains 15% N (0.15 x 75 = 11.25). If 1 ounce of calcium nitrate is dissolved in 100 gallons of water, the solution will contain approximately 11.25 ppm N.

II. To calculate the number of ounces of material required to make up a desired ppm concentrate, solve for C:

$$C = \frac{\text{Desired ppm conc.}}{B}$$

Where:
B = ppm contained in 1 ounce of the material in 100 gallons of water (from above)
C = number of ounces of material to add to 100 gallons of water to achieve the desired concentration

Example: To make up a 250 ppm solution of calcium nitrate, first multiply the AI x 75 (.15 x 75 = 11.25). Next divide the desired concentration by 11.25 (250 ÷ 11.25 = 22). To make up a 250 ppm solution of calcium nitrate, you would add 22 ounces to 100 gallons of water.

Practical Examples

I. Make up a nutrient solution of 300 ppm N. Half of the N to be supplied by ammonium nitrate. (Ammonium nitrate = 33% N, Calcium nitrate = 15% N).

$$.33 \times 75 = 29.75$$
$$.14 \times 75 = 11.25$$
$$\frac{150}{29.75} = 5 \qquad \frac{150}{11.25} = 13$$

To make up a 300 ppm nutrient solution of 1/2 ammonium nitrate and 1/2 calcium nitrate, dilute 5 and 13 ounces respectively in 100 gallons of water.

II. Make up a 200 ppm solution of ammonium nitrate for use with a 1:100 proportion.

$$.33 \times 75 = 29.75$$

$\dfrac{200}{29.75}$ = 7 ounces of ammonium nitrate/ 100 gallons water.

7×100 = 700 to concentrate the solution 100 times so that when proportioned it will distribute a 200 ppm solution.

To make up a 200 ppm solution of ammonium nitrate for use with a 1:100 proportion, dilute 700 ounces of material in 100 gallons of water.

Language — Communicate — Understand

The dealer, the salesman, and the grower all need to enlarge their vocabulary to include all these terms of biological farming.

Grower language consists of the words, their pronunciation, and the methods of combining them, using them, understanding them as they relate to today's biological agriculture.

Communicating implies making common to all what one presently possesses. The purpose of this publication is to share with the reader what I have learned about seaweed and plant growth over the past thirty years.

Hopefully, the reader will achieve a grasp of the nature, significance, and understanding of how plants grow.

CHAPTER 12
There Is a Time for All Things

This time, like all times, is a very good one if we but know what to do with it (Emerson).

There's a Time for All Things

In Shakespeare's "The Comedy of Errors," he states that "There's a time for all things." How true that statement is in relationship to today's world of biological agriculture. The process by which soils were made took millions of years, but mankind has been able to alter and change this in a matter of a few years. Appropriate today, then, is, "There's a time for all things," and today it is biological — agriculture to help Mother Nature salvage our soils and restore good healthy conditions for plant growth. This will take time and money, but the rewards will be great.

Nature should be watched and observed, a little something done each day, but do not expect it all to happen at once.

Time

Did you ever slow down in life enough to evaluate and appreciate time? Another way of asking the question: have you ever taken time out to think how often the word **"time"** enters your lifetime activities?

In the morning a grower gets up early to take advantage of the daytime hours. It is in these daytime periods that plants are using essential elements, micronutrients, water, growth regulators, and, of course, light.

Regulate is directly associated with time; for regulate means to fix or adjust the time, amount, degree, or rate of something. The words fix, adjust, amount, and rate are very important to the grower. For example, the grower decides to use a plant growth regulator, he must first decide the **amount** to use, next he will determine the rate per acre or per field. This decided, he begins to fix the sprayer, perhaps we should say he will **adjust** the calibration to **fix** the **amount,** of actual material to apply.

Don't forget **time** is important if you expect to get your just return on your investment. **Timing** of application must be tailored to the crop with strict compliance with label directions. Oh, yes, the **time** of day is very important in considering when to apply the foliar spray. Early morning, late afternoon, early evening are ideal **times** to apply seaweed extracts. Avoid midday to midafternoon, the leaves may be wilted, plant is low on nutrients, temperature too high for the plant to efficiently take in the much needed nutrients and growth regulators.

Remember there is a time to spray and a time not to spray.

Time may be considered as the measured or measurable period during which an action, **process,** or condition exists or continues.

Perhaps the most common reason for the failure of plant growth regulators, micronutrients, and foliar sprays is the fact that growers do not apply the material at the most critical time. Therefore, when the label states — **"Time** of application" — bloom time, etc., that is the time that research has proven that the particular **plant process** is most active and in critical need of the growth promoting substance.

A good example is found in the summary statements of M. Povolny reporting on his investigation of the effectiveness of an extract of seaweed **(ASCOPHYLLUM nodosum)** on the yield and quality of cucumbers for pickling.

Povolny states, "Water sprays of a seaweed concentration of 0.04 percent applied **at the time** of the fruiting of cucumbers in weekly intervals distinctly increased the yields of fruits by 41.8 percent." The **process** involved is that of changing from a vegetative plant to a reproductive plant. At this time it is very important that the certain micronutrients are available to the plant to assist in the fertility of the male and female organs of the cucumber flowers.

As stated earlier, time is considered as the measurable period during which a condition exists or continues. The importance of **time** of spray applications is further substantiated by Povolny. He reports, "The marketing quality of stored cucumbers from sprayed plants was prolonged for a **time** period by 14 to 21 days. They showed a superior resistance to softening and rotting compared with fruits from the control plants."

Jeanne W. Price, widely known exhibition judge and member of the American Iris Society, through her research program, said that it has been established that by applying granular seaweed at planting time, again prior to bloom time, and spraying with seaweed concentrate thereafter at two week time intervals, iris grow well and become more productive.

Price concluded that seaweed **(ASCOPHYLLUM nodosum)** used exclusively in her tests, shows its value in the growing of iris by their vigorous, clean, healthy growth, increased blossoms, their extended blooming season (a longer

time than ever before), their enriched colors, and rapid and high increase in size.

T. el Kobbia of Ain Shama University, Cairo, Egypt, reported that from the research results of his experiments with wheat and soybean that the micronutrients in seaweed concentrate are adequate for normal deficiency problems under Egyptian conditions. He points out that the interesting yield increase of 43% obtained by seaweed concentrate might also suggest that plant hormones along with micronutrients content of seaweed concentrate might be responsible for the high beneficial effect of these foliar application on cotton yield in Egypt. T. el Kobbia, Professor of Plant Nutrition, concluded by saying, "We believe that the **time** and/or the frequency of seaweed applications are of great importance for beneficial effect on the yield of soybean."

Disease

Several scientists have noted that seaweed concentrates, in view of their cytokinin content, may effect the resistance of plants to disease. They state that while not eliminating the infestation itself, the applied cytokinins apparently allow the plant **time** to increase its resistance to the disease.

Dekker (1963) noted that applied kinetin inhibited powdery mildew on cucumbers. Senn (1961) observed that seaweed concentrate applied to cantaloupe plants reduced the severity of powdery mildew on leaves. Wang (1961) stated that applications of kinetin decreased the occurrence of stem rust in wheat.

Some researchers think that, because laboratory trials failed to disclose any direct fungicidal or fungicidal principle effective against **Botrytis** organism, that late applications of seaweed concentrate before fruiting began on strawberries also could have no direct prophylactic effect. They conclude that there is evidence that the effect of seaweed extracts in reducing the incidence of disease lies in the changes it produces in the plant itself. These may be small, structural changes such as increased thickness of cuticle, or cell walls, or reduction in stomatal size; or they may be complex physiological changes affecting a whole variety of internal defense mechanisms, including alterations in the pH of the cell sap.

There is no limit to the numerous ways in which seaweed may be used in efficient plant growth and development.

Time flies, we say, alas-alak, time stays — we go!

To every THING there is a season, and a time to every purpose under the heaven: A time to be born, and a time to die; a time to plant, and a time to pluck up that which is planted (Ecclesiastes 3:1-2).

Things

There is a time for some THINGS, and a time for all THINGS, a time for great THINGS, and a time for small THINGS (Cervantes — Don Quixote).

Ideas are funny little **things.** They won't work unless you do.

The news media saturates us with profound statements — today's agriculture requires high yields of good quality.

However matter-of-fact this idea may be, growers are finding it harder to accomplish this each year. More and more N-P-K each year per acre is just not getting the job done.

Agriculture today is a challenging business enterprise. Farming has always been a tough profession. Farmers have been told that high yields mean high profits. This idea just hasn't worked out; for the business side of farming hasn't kept up with the production side.

Growers are seeing more soil problems each year. One of these problems is soil compaction, which in turn reduces soil aeration. Another problem is the poor decomposition of crop residue. Growers complain of sick or sterile soil resulting from too much herbicides. All of these problems lead to a reduction of beneficial soil microorganisms. In some areas nutrient elements have become complexed to a point where they are no longer available to the plants.

The number of plant growth regulators — hormones — available to growers to use in many ways is almost unlimited. Trade magazines report that agrichemical interests are investing heavily in plant growth regulator research. It is estimated that the plant growth regulator market will be as large or larger than the pesticide market by the year 2000. One report stated that as many as 20,000 synthetic compounds are annually screened for plant growth regulator activity by the chemical industry.

Things to Consider

Many things go wrong when growers use new products and new technology. Seaweed extracts may not be new, but new technology has resulted in highly concentrated products that require specific techniques in their use.

Plant growth regulators are used in very small quantities, generally in the range of parts per million or even parts per billion. Even some of the old N-P-K advocates that preach thousands of pounds per acre, have trouble understanding pints and quarts that are recommended. Minute amounts of active ingredient are applied, which necessitates that spray equipment be correctly calibrated and in good working order.

Different crops have different physiological activities. There is no "across the board" recommendation. Timing of application is very exacting and must be initiated to do what the plant process utilized is intended; for example, cell formation (growth), flowering, healing of cut surfaces (alfalfa), root formation and growth, fruit set stress, and prolonging shelf life.

Concentration is vital to success in the use of seaweed extracts. The old story, "When all else fails, read the instructions," is very true when using seaweed extracts. Research has shown the best time and the best concentration to apply seaweed extracts.

Weather conditions cannot be controlled, and every grower knows that sometimes weather is your worst enemy. Environmental conditions, such as moisture, temperature, and relative humidity actually determine plant activity. Time of day is very important in considering when to apply seaweed extracts. Always avoid middle of the day when planning spray schedules.

Beware

The grower should not be pressured by a scheming salesperson with dollar signs in his eyes into buying some product they really don't need or can't afford, and should not be taken in by any number of tricks the salesperson might employ. There are, indeed, dishonest seaweed salespersons, as there are dishonest people in any walk of life, but there are also many seaweed salespersons who do their best to help a grower make the wisest, and best, most practical purchase for his or her needs. These dealers, distributors, and salespersons have earned the respect of fellow growers as well as the general public.

The Team

It takes team work to be a champion. N-P-K alone will not solve the problem. Seaweed extracts alone will not solve the problem. Micronutrients alone are not the answer. The team combination will make champions of crop plants.

Seaweed extracts may be applied by spraying with airplane, ground sprayer, through sprinkler systems, or by furrow or flood irrigation. These products may be applied with most liquid fertilizers, herbicides, and pesticides. Always read and follow very carefully all instructions printed on the label.

Join the championship team, keep up with the new advancements taking place in agricultural biotechnology. Nothing will ever replace common sense, hard work, and an open mind in the most rewarding of all professions — **AGRICULTURE.**

All times are good times. This time is only the beginning of new technologies. We need not mistrust them nor be reluctant to relinquish old ways. We must all try to understand the beneficial effects and evaluate every alternative.

Time and Research March On

Dr. Wayne Temple and coworkers at the University of British Columbia (1986) have just completed an outstanding experiment evaluating the potential uses of kelp as a foliar spray. Their publication should now be available.

Dr. T. el Kobbia, Professor of Plant Nutrition, Faculty of Agriculture, Ain Shams University, Shubra El Kheima, Cairo, Egypt, is conducting research with Norwegian **ASCOPHYLLUM nodosum** in production of cotton, wheat, grapes, soybeans, green beans, and citrus.

Drs. B.C. Featonby-Smith and J. van Staden of the Department of Botany, University of Natal, Pietermaritzburg, 3200, Republic of South Africa, continue to do excellent research with seaweed concentrates on the growth of a number of vegetable crops.

Dr. Mordechai Gersawi of the Department of Biology, Ben-Gurion University of the Negev, Beersh Eva., Israel, has conducted outstanding research on cytokinin-stimulated translocation in isolated bean leaves.

Dr. C.L. Young, Department of Physical Chemistry, University of Melbourne, Parkville, Victoria 3052, Australia, continues to publish on his research with **ASCOPHYLLUM nodosum**.

Dr. G. Blunden and associates, School of Pharmacy, Portsmouth Polytechnic, King Henry I Street, Portsmouth, PO1 2DZ UK, continue to make valuable research contributions on seaweed extracts. Drs. Alhaji S. Jeng and Einar Vigerust of the Department of Soil Fertility and Management, Agricultural University of Norway, N-1432 ASNLH, Norway, are researching the effects of kelp-meal on some physical properties of the soil.

Drs. R.A. Cline and Neil W. Miles, Ministry of Agriculture and Food, Horticultural Research Institute of Ontario, Canada LORZEO have research underway evaluating the influence of Norwegian **ASCOPHYLLUM nodosum** on yield and quality of apples, grapes, peaches, and strawberries. Research evaluations of Norwegian **ASCOPHYLLUM nodosum** influence on nematodes attacking tomato and tobacco are also underway in Ontario.

Numerous trials are in progress in Islamabad and Karachi, Pakistan, being conducted by the Pakistan Agricultural Research Council with Norwegian **ASCOPHYLLUM nodosum**.

Field evaluation of Norwegian **ASCOPHYLLUM nodosum** liquid seaweed extract as a seed dressing and foliar treatment of dryland winter wheat **(Triticum aestiuum** L. var. Stephens) in the United States inland Pacific northwest is now being made.

Dr. A.C. Tarjan, Professor and Nematologist, Department of Entomology and Nematology, University of Florida, Gainesville, Florida 32611 U.S.A., has exciting research under way evaluating various seaweed products on nematode control in numerous horticultural crops.

Metting and Pyne (1986) state that numerous plant growth regulators have been discovered as components of most groups of microorganisms (especially fungi) and of seaweeds. They further state that "the ability to easily control growth conditions in a bioreactor together with the demonstrated biochemical diversity of microalgae suggests that these organisms will become the focus of greater screening and selection efforts for biologically active compounds."

Metting, et al. (1986), report that the larger share of the market for seaweed products is occupied by sales for use in glass house and other intensive horticultural operations. They add that should seaweed products be shown to perform consistently with agronomic cultivation of cereals by increasing herbicide use efficiency, by promoting yields or by both, however, use on an agricultural scale is possible.

This listing is not exhaustive and many projects may have been omitted; however, the manuscript has been designed to be revised as necessary.

New Product Development

Research and development efforts by commercial seaweed processors have resulted in recent releases of new products designed for specific uses. Norwegian manufacturers are cold processing **ASCOPHYLLUM nodosum** to be used as a basis for liquid or powder specific formulations.

Product **AG** is a cold processed extract powder retaining all components of the seaweed, and product **BB** is a cold processed extract liquid retaining all components of the seaweed at a low pH. Both product uses are foliar spray, agriculture, commercial growers, nurseries, root dipping, soil application, land reclamation, and as a transplanting starter.

The following tables illustrate the advantages of the new products over former processing methods.

Table 1

Bean growth was greater in the media containing seaweed extracts than in the medium containing 1 mg/1 Kinetin.

Supplementation to basal medium	Increase in fresh weight over basal medium (%)	Increase in fresh weight over Kinetin (%)
a. 1 mg/1 Kinetin	21.8	—
b. 0.1% w/v comm AF (1 g/l)	22.7	0.9
c. 0.1% w/v AG (1 g/l)	18.0	6.2
d. 0.125% w/v AG (1 g/l)	27.7	5.9

Table 2

Radish test comparing new **ASCOPHYLLUM** products with control and former processed product.

Foliar Spray	Average Height (%)	Average Weight
Water (control)	100	100
Comm AF	102.6	102.3
AG	109.8	100.6
BB	121.7	104.8

Table 3

Response of transplanted palm trees to root applications of new cold extract **ASCOPHYLLUM** product.

	Leaves elongation (cm)			
Treatment	30 days	60 days	80 days	Total
Control (water)	0.5	5.0	3.5	9.0
3% AG solution	2.5	5.5	3.5	11.5

Hydro-Seeding

The development of the hydro-seeder and new formulations of seaweed products has enabled the seeder to sow grass on heretofore impossible terrain. Seaweed components contain sodium alginates and alginic acid, both of which improve water holding capacities. The seaweed water soluble materials contain growth promoting substances as well as a large number of micronutrients which correct soil nutrient deficiencies as well as increase the formation of humus.

Tree and Shrub Transplanting

Recent research of moisture content of plants during transplanting and the development of protective root dips of seaweed has shown great reduction in loss of young transplants.

The roots of trees or shrubs are dipped in seaweed solutions. After allowing the excess solution to drain away the plants may be stored or planted. These formulations are widely used in Europe with great success.

Accuse not nature! She hath done her part; do thou but thine (Milton — *Paradise Lost*).

Timely Research Publications

1. Temple, W.D., R.A. Radley, A.A. Bomke. 1986. The potential use of the kelp **MACROCYSTIS integrifolia,** as a foliar spray in agricultural crop production. Doctoral Thesis, University of British Columbia, Vancouver, British Columbia.
2. Metting, B. and J.W. Pyne. 1986. Biologically active compounds from microalgae. Enzyme Microb. Technol., 1986, Vol. 8. July, 385.
3. Metting, B., W.R. Rayburn, and P.A. Reynaud. 1986. Algae and agriculture. Algae and human affairs. Edited by C.A. Lembi and R.A. Waaland for Cambridge Univ. Press, Cambridge, England.
4. Raviv, M. The effect of seaweed concentrate on "bottom break" formation of roses. Hassadeh, 1985. 65,5 952-954.
5. Tay, S.A.B., J.K. MacLeod, L.M.S. Palni, D.S. Letham. Detection of cytokinins in a seaweed extract. Phytochemistry (OXF), 24(11), 1985. 2611-2614.
6. Featonby-Smith, B.C., J. Van Staden. The effect of seaweed concentrate and fertilizer on growth and the endogenous cytokinin content of Phaseolus vulgaris. S. Africal J. of Botany, 1984. 3,6, 375-379.
7. Dwelle, R.B., P.J. Hurley. The effects of foliar application of cytokinins on potato yields in southeastern Idaho. Am. Potato J., 1984. 61,5, 293-299.
8. Hare, R.C. Stimulation of early height growth in longleaf pine with growth regulators. Canadian J. of Forest Rsch., 1984. 14,3, 459-462.
9. Marcelle, R.D. Effects of GA3, BA and growth retardants on fruit set in the pear cultivar "Doyenne du Comice." Acta Horticulturae, 1984. No. 149, 225-229.
10. Nelson, W.R., J. Van Staden. Growth regulation of wheat triticum-aestivum by seaweed **ECKLONIA maxima** concentrate application. Annual Meeting Am. Soc. of Plant Physiologists, Davis, CA, USA, Aug. 12-17, 1984. Plant Physiol. 75 (Suppl. 1). 1984. 132.

11. Lane, D.J., A.R. Langille. Influence of plant growth stage and concentration of Cytex and kinetin applications on tuber yields of two potato cultivars. HortScience, 1984. 19:4, 582-3.
12. Featonby-Smith, B.C., J. Van Staden. The effect of seaweed concentrate and fertilizer on the growth of Beta vulgaris. Zeitschrift fur Pflanzenphysiologie, 1983. 112,2, 155-162.
13. Tarjan, A.C., J.J. Frederick. Comparative effects of kelp preparations and ethoprop on nematode-infected bermudagrass. Nematropica, 1983. 13,1, 55-62.
14. Featonby-Smith, B.C., J. Van Staden. The effect of seaweed concentrate on the growth of tomato lycopersicon-esculentum plants in nematode infested soil. Sci. Hortic. (AMST) 20(2), 1983. 137-146.
15. Bryan, H.H., F.A. Maas, Mary Sherry. Fluid drilling of pregerminated pepper seed. Proc. Fla. State Hortic. Soc., 1983. 95, 353-356.
16. Wilczek, C.A., T.J. Ng. Promotion of seed germination in table beet by an aqueous seaweed extract. HortScience, 1982. 17,4. 629-630.
17. Kingman, A.R., J. Moore. Isolation purification and quantitation of several growth regulating substances in **ASCOPHYLLUM nodosum** phaeophyta. Bot. Mar. 25(4), 1982. 149-154.
18. Dybing, C.D., C. Lay. Oil and protein in field crops treated with morphactins and other growth regulators for senescence delay. Crop Sci., 1982. 22:5, 1054-8.
19. Luanratana, O., W.J. Griffin. The effect of a seaweed extract on the alkaloid variation in a commercial plantation of a Duboisia hybrid. J. Nat. Prod., 1982. 45:3, 270-1.
20. Tarjan, A.C., J.J. Frederick. Reaction of nematode-infected centipedegrass turf to pesticidal and non-pesticidal treatments. Proc. Fla. Hort. Soc., 1981, publ. 1982. 94, 225-227.
21. Sciuto, S., R. Chillemi, M. Piattelli, G. Puglisi. Gibberellin-like and cytokinin-like activities in marine algae from central Mediterranean. Boll. Soc. Ital. Biol. Sper 57(15), 1981. 1590-1595.

22. Dybing, C.D., C. Lay. Yields and yield components of flax, soybean, wheat and oats treated with morphactins and other growth regulators for senescence delay. Crop Sci., 1981. 21:6, 904-8.
23. Dybing, C.D., C. Lay. Field evaluations of morphactins and other growth regulators for senescence delay of flax, soybean, wheat, and oats. Crop Sci., 1981. 21:6, 879-84.
24. Shanks, J.B. Promotion of greenhouse cut roses with GA4,7 and benzyladenine. Proc. Plant Growth Regul. Soc. Am., 1981. Vol. 8, 224-30.
25. Sharifihosseini, H. Influence of cytex (a growth regulating compound), bed form and irrigation on fruiting and vegetative characteristics of "Guardian" and "Sunrise" strawberry (Fragaria x annanasa Duch.) cultivars. Dissertation Abstract, 1981.
26. Ketring, D.L., A.M. Schubert. Reproduction of peanuts treated with a cytokinin-containing preparation. Agron. J., 1981. 73:2, 350-2.
27. Abetz, P. Seaweed extracts: have they a place in Australian agriculture or horticulture? J. Australian Inst. of Ag. Science, 1980. 46-1, 23-29.
28. Criley, R.A. Stimulating lateral bud break on Dracaena. Plant Propagator, 1980. 26:2, 3-5.
29. Kentzer, T., R. Synak, K. Burkiewicz, A. Banas. Cytokinin-like activity in sea water and fucus-vesiculosus. Biol. Plant (Prague) 1980. 22(3), 218-225.
30. Luanratana, O., W.J. Griffin. Cultivation of a Duboisia hybrid. Part B. Alkaloid variation in a commercial plantation: effects of seasonal change, soil fertility, and cytokinins. J. Nat. Prod., 1980. 43:5, 552-8.
31. Blunden, G., P.B. Wildgoose, F.E. Nicholson. The effects of aqucous seaweed extract on sugar beet. Botanica Marina, 1979. Vol. 22, 539-541.
32. Hurtt, W., R.B. Taylorson. Greenhouse studies of chemical effects on wild oat seed shedding and germination. U.S. Dept. Agric., ARS, Frederick, Md. 21701 U.S.A., 1978, Vol. 32, 114.

33. Borowczak, E., T. Kentzer, B. Potulska-Klein. Effect of gibberellin and kinetin on the regeneration ability of fucus-vesiculosus. Biol. Plant (Prague), 1977. 19:6, 405-412.
34. Kingman, A.R., T.L. Senn. Bioassay systems to test for plant growth hormones in extracts of **ASCOPHYLLUM nodosum**. J. Phycol. 13, 1977. 36.
35. Fries, D. Phenyl acetic acids native growth regulators in seaweeds. J. Phycol. 13, 1977. 23.
36. Blunden, G., P.B. Wildgoose. The effects of aqueous seaweed extract and kinetin on potato yields. J. Sci. of Food & Agric., 1977. 28:2, 121-125.
37. Dybing, C.D. Delayed senescence of flax treated with morphactins or anti-auxins. J. Proc. Plant Growth Regul. Work. Group, 1977. Vol. 4, 207-210.
38. Augier, H. Phyto hormones of algae. Part 2. Physiological study. Ann. Sci. Nat. Bot. Biol. Veg. 1974. 15:2, 119-180.
39. Augier, H. Plant Hormones in Algae. Part 1. Biochemical Study. Ann. Sci. Nat. Bot. Biol. Veg. 15:1, 1-64.
40. Augier, H., H. Harada. Contribution to the study of endogenous cytokinins in algae. Tethys., 1973. 5:1, 81-93.
41. Pedersen, M. Identification of a cytokinin 6-3 methyl-2 butenylamino purine in sea water and the effect of cytokinins on brown algae. Physiol. Plant, 1973. 28:1, 101-105.
42. Augier, H., H. Harada. Presence of hormones of the cytokinin type in the thallus of marine algae. Seances Acad. Sci. Ser. D Sci. Nat. 275 (16) 1972, 1765-1768.
43. Pedersen, M., G. Fridborg. Cytokinin-like activity in sea water from the fucus-**ASCOPHYLLUM** zone. Exper. Suppl. (Basel) 1972. 28:1, 111-112.
44. Jennings, R.C., W.J. Broughton, A.J. McComb. Effect of kinetin on the phycoerythrin and chlorophyll content of a red alga. Phytochemistry, 1972. 11:6, 1937-1943.
45. Jennings, R.C. Cytokinins as endogenous growth regulators in the algae **ECKLOINIA**-radiata phaeophyta and hypnea-musciformis rhodophyta. Aust. J. Biol. Sci., 1969, 22:3, 621-627.

Nature reveals her secrets reluctantly. Research and development will discover new products and find answers to how plants grow.

Author Index

A

Abetz, P. 5-3, 6; 7-5, 6; 12-15
Aitken, J.B. 3-1, 5, 6, 16; 5-5, 6; 7-4, 6
Alexander, M. 5-6
Armstrong, D.J. 5-6
Augier, H. 3-2, 9, 10, 11, 16; 12-16
Austin, M.D. 10-34

B

Badizadegan, M. 10-34
Baker, R.C. 3-3
Bentley, J.A. 3-8, 10, 14, 16
Biddington, N.L. 10-34
Blunden, G. 3-1, 2, 5, 7, 8, 13, 16; 5-5, 6; 10-22
Boney, A.D. 3-12
Bonner, J. 10-11
Booth, E. 3-1, 17; 5-5, 6; 10-10, 11, 34
Borowczak, E. 12-14
Brian, K.R. 3-8, 11, 12, 17; 5-7
Briner, G.P. 5-7
Brueske, C.H. 5-7
Bryan, H.H. 12-14
Button, E.F. 5-7

C

Cathey, H.M. 10-34
Chalopin, M.C. 3-17
Clayton, M.N. 2-7
Cline, R.A. 12-9
Cook, R.L. 10-34
Crawford, J.H. 1-6
Criley, R.A. 12-15

D

Darrah, C.H. 3-14, 15, 17
Dekker, J. 12-4
Derting, C.W. 10-5
Driggers, B.F. 5-7; 10-14, 34
Dropkin, V.H. 5-7; 10-34
Duff, R.B. 5-7
Dwelle, R.B. 12-13
Dybing, C.D. 12-14

E

Eidt, D.C. 9-5, 8
Esashi, Y. 10-34

F

Falgout, R.N. 3-2, 17
Featonby, S. 5-3; 10-34; 12-8, 13

Fergusson, M. 5-7
Fogg, G.E. 5-7
Fox, D. 1-6
Fox, J.E. 5-7
Franki, R.I. 3-5, 17; 7-4
Fries, L. 3-4, 17; 12-15

G

Gagon, J.D. 5-7
Gambrell, C.W. 1-7
Garman, P. 10-12, 34
Gersani, M. 12-9
Goh, K.M. 5-5, 7
Gubensek, F. 3-2

H

Hall, F.R. 3-14, 15
Hare, R.C. 12-13
Harley, J.L. 5-8
Hiron, R.W. 9-5, 9
Hurley, P.J. 12-13
Hurtt, W. 12-15
Hussian, A. 3-6, 12, 17

I

Ikawa, T. 8-8
Itai, C. 9-5, 8

K

Kelly, J.W. 11-6
Kentzer, T. 5-8; 12-15
Ketring, D.L. 12-14
Kettellapper, H.J. 10-10, 11
Khaleafa, A.F. 5-8; 10-34
Kingman, A.R. 3-1, 7, 18; 8-4, 8; 10-6
Klambt, D. 5-8
Kobbia, T. El 12-4, 8

Kochba, J. 5-8; 10-34
Kuribayashi, T. 10-34

L

Lane, D.J. 12-13
Langille, A.R. 12-12
Letham, D.S. 2-7
Levitt, J. 9-2, 8
Lewis, J.A. 1-8
Lewis, L.N. 1-2
Little, C.H.A. 9-8
Luanratana, O. 12-15
Lynn, L.B. 3-5, 18; 7-4, 6

M

MacLeod, J.K. 3-15; 8-8; 12-11
McDonough, W. 10-34
Makino, H. 10-34
Martin, J.A. 1-6, 7
Marcelle, R.D. 12-13
Marrucci, P.E. 10-14
Marth, P.C. 10-11, 34
Mateus, H. 3-3, 18
Metcalf, R. 1-2
Metting, B. 12-10
Miles, N.W. 12-9
Miller, C.O. 5-8
Milton, R.F. 5-8
Moore, J. 10-6
Mowat, J.A. 3-10, 18; 10-6
Munda, M. 3-2, 3, 18

N

Nelson, W.R. 12-12
Nicholson, F.E. 3-8
Nisizawa, K. 3-19

O

Offermans, C.N. 3-18
Ohrstrom, P.B. iv

P

Palni, L.M.S. 8-8
Paracer, S. 10-14, 17
Pedersen, M. 3-11, 13, 18; 12-14
Povolny, M. 3-1, 6, 18; 5-5, 8; 12-3
Price, J.W. 12-3
Puls, E.E., Jr. 10-35

R

Railton, I.D. 9-5, 8
Ramshaw, D. 5-8
Raviv, M. 12-11
Regenstein, J.M. 3-3
Reid, D.M. 9-5, 8
Reid, S.M. 3-10
Reynolds, T. 10-35
Rikin, A. 10-12

S

Sawhney, R. 5-9; 10-35
Sciuto, S. 12-14
Senn, T.L. 3-1, 4, 5, 6; 5-8; 8-4
Shanks, J.B. 12-14
Sharifihosseini, H. 12-14
Sharples, G.C. 10-35
Silvalingam, P.M. 3-19
Skelton, B.J. 5-9
Slade, D.A. 5-9
Spaull, V.W. 5-9
Stephenson, W.A. 5-5, 9

T

Tarjan, A.C. 10-17; 12-9, 13
Tay, S.A.B. 8-8; 12-13
Temple, W. 12-13
Terriere, L.C. 5-9
Turner, T.D. 3-17

U

Upper, C.D. 3-15

V

Van Staden, J. 5-3, 9; 12-8
Vigerust, E. 12-9

W

Wang, D. 5-9
Weidman, R.W. 5-3; 9-8
Whitehead, A.G. 5-9
Wilczek, C.A. 12-14
Widdowson, J.P. 5-9
Wildgoose, P.B. 3-8
Williams, D.C. 5-9
Wright, S.T.C. 9-5, 9

Y

Young, C.L. 5-3; 7-5; 8-8; 10-6

Subject Index

A

ABA 2-4; 9-5
Abscissic Acid 2-4; 10-12
Adenine 2-4; 8-9
Adenosine 2-4
Alfalfa 10-26
Algae 2-1
Alginic Acid 2-5; 3-7
Apical Control 11-4
Apple 8-4; 10-5, 12
Apricot 10-7
Ascophyllum 1-2, 3; 2-1, 3; 3-1; 5-4
Auxin 2-3; 8-3, 5; 10-22

B

Banana 10-12
Basics 10-2
Beta vulgaris 5-3; 9-3
Beware 12-7
Bioassay Tests 3-7; 8-4
Biochemical 7-1; 11-1
Biological 8-10; 10-1
Bioliving 11-1
Biophysical 11-2
Boron 7-3
Botrytis 10-10; 12-4
Bromophenols 10-17

C

Camellia 10-29
Carnations 10-7
Carrot 3-13
Cauliflower 5-4
Celery 10-4
Chelating 4-14; 10-14; 11-3
Chondrus 2-2
Chlorophyll 3-14; 4-1
Chlorophyta 3-2, 12
Cherries 10-7
Chrysanthemum 3-1
Citrus 3-5
Clover 5-5
Cold Hardiness 10-10
Communication 11-8
Compatibility 2-6, 8
Crop Resistance 10-11, 12
Cucumbers 9-6; 12-3
Cytokinin 1-4; 2-4; 3-11; 8-5, 7, 8, 9

D

Deficiency 4-14
Diseases 12-4
Drought 9-4
Durvillea 1-3; 2-7; 5-5; 8-8

E

Ecklonia 1-3; 5-3
Enteromorpha 3-12
Environment 9-2, 4
Environmental Stress 9-4
Enzymes 8-1
Essential Elements 6-3; 7-2; 11-2
Extracts 4-4

F

Flowering 6-2; 9-7; 10-6
Flowers 4-15; 9-7
Frost Damage 9-7; 10-10
Fruit Set 10-7, 13
Fucaceae 1-3; 2-1; 3-8
Fungi 10-10

G

Geranium 10-33
Gibberellin 1-4; 3-9; 8-3, 5, 7
Gigartinaceae 1-3
GLC 1-4; 3-15; 8-9
Gladioli 3-5
Grapes 10-8
Grasses 3-14; 10-6, 27
Grower Comments 10-22
Grower Usage 10-22
Growth 4-2, 16; 5-1
Growth Hormones 8-2
Growth Promoters 1-3; 2-3; 4-21
Growth Regulators 8-1; 11-4
Gymnodinum 3-10

H

Herbicides 10-31, 32
Holly (Ilex) 10-6
Hoplolamus 10-17
Hormones 5-2; 6-4; 8-2

I

IAA 1-4; 2-4
Ilex 4-5; 10-6
Insects 10-14
Iris 12-3
Iron 3-6; 4-9; 7-3

L

Language 11-1, 8
Laminaria 2-2; 3-5, 8, 11
Leaf 4-12
Limes 10-22
Limiting Factor 6-3

M

Macrocystis 2-2; 3-3
Magnesium 3-5
Manganese 7-3
Mannitol 2-5; 4-14
Mannose 2-5
Maturity 6-4
Meloidogyne 10-17
Melon 10-10
Micronutrients 5-2; 7-1, 2, 3; 10-3
Mildew 10-10
Mineral Nutrition 7-2; 11-3

N

NAA 10-22
Nandina 4-5
Nematodes 3-14; 10-16; 12-9
Nicotine 10-27; 12-9

O

Okra 10-32
Onion 4-8
Oranges 3-5

Organic Acids 2-4
Osmotic Stress 9-1

P

Pallisade Cells 4-13
Panax-Ginseng 10-4
Peaches 3-6; 10-18
Peppers 8-10; 10-22, 24
Periodisms 3-10
Phaeodactylum 3-2, 10
Phaeophyta 1-8
Phenyl Acetate 3-4, 9
Photosynthesis 7-1; 11-1
Pinus 4-4, 6
Pisum 4-4
Plant Foods 11-5
Plum 10-7
Potatoes 3-8; 10-6, 29
ppm-ppb 2-3; 11-5, 6

R

Radish Tissue Test 3-11, 13
Red Spider 10-14
Regeneration 3-10
Reproductive Phase 6-2
Research Grants 1-5; 10-23
Rhododendron 10-5
Rhodophyta 3-2, 12
Root 4-8; 10-5

S

Sargassum 3-2
Seed 4-3, 8; 10-4, 25
Seed Physiology 4-3; 9-3
Senescence 10-21
Sexual Mechanisms 4-15
Shelf Life 10-18, 22
Soybeans 3-13; 12-4
Spatoglossum 10-17

Stem 4-11
Stomates 4-13
Strawberry 9-8; 10-9, 10
Stress 9-1, 2
Sugars 2-4, 5; 3-8
Sunflower 9-5; 10-6
Sweet Corn 10-6
Sweet Potato 10-31
Symposia 1-4

T

Tanning 10-17
Temperature 9-5
Terpenoids 10-17
The Team 12-7
Time 12-1
Tomatoes 5-4; 10-10
Transpiration 11-2
Tropisms 3-10; 8-1

V

Vegetative Phase 6-1

W

Wheat 10-7, 30; 12-4, 9

Z

Zeatin 2-4, 7
Zinc 7-3
Zinnia 4-4